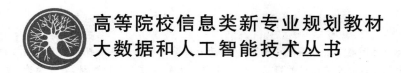

高等院校信息类新专业规划教材
大数据和人工智能技术丛书

Hadoop 大数据平台开发运维实训

主　编　余　挺
副主编　张　浩　李　超

北京邮电大学出版社
www.buptpress.com

内容简介

大数据时代的到来，迫切需要高校及时建立大数据技术课程体系，为社会培养和输送一大批具备大数据专业素养的高级人才，满足社会对大数据人才日益旺盛的需求。本书定位为大数据技术入门教材，旨在为读者搭建起通向"大数据知识空间"的桥梁。本书将系统地梳理总结 Apache Hadoop 大数据相关技术，介绍大数据存储、并行计算、数据处理等内容，帮助读者形成对大数据知识体系及其应用领域的轮廓性认识，为读者在大数据领域进行更深入的学习和研究奠定基础、指明方向。在本书的基础上，感兴趣的读者可以通过其他诸如《大数据技术原理及应用》《Hadoop 权威指南》等工具书，深入学习和实践大数据相关技术。

本书可作为高等院校计算机、信息管理等相关专业的大数据课程教材，也可供相关技术人员参考、学习、培训之用。

图书在版编目(CIP)数据

Hadoop 大数据平台开发运维实训 / 余挺主编． －－ 北京：北京邮电大学出版社，2022.1(2024.7 重印)
ISBN 978-7-5635-6584-9

Ⅰ．①H… Ⅱ．①余… Ⅲ．①数据处理软件—教材 Ⅳ．①TP274

中国版本图书馆 CIP 数据核字(2021)第 261624 号

策划编辑：刘纳新　姚　顺　　责任编辑：徐振华　米文秋　　封面设计：七星博纳

出版发行：北京邮电大学出版社
社　　址：北京市海淀区西土城路 10 号
邮政编码：100876
发 行 部：电话：010-62282185　　传真：010-62283578
E-mail：publish@bupt.edu.cn
经　　销：各地新华书店
印　　刷：保定市中画美凯印刷有限公司
开　　本：787 mm×1 092 mm　1/16
印　　张：13.5
字　　数：351 千字
版　　次：2022 年 1 月第 1 版
印　　次：2024 年 7 月第 3 次印刷

ISBN 978-7-5635-6584-9　　　　　　　　　　　　　　　　　定　价：38.00 元

・如有印装质量问题，请与北京邮电大学出版社发行部联系・

前　言

进入 2012 年后，大数据（Big Data）一词越来越多地被提及，人们用它来描述和定义信息爆炸时代产生的海量数据，并命名与之相关的技术发展与创新。

随着云时代的来临，大数据吸引了越来越多的关注。数据正在迅速膨胀并变大，它决定着企业的未来发展，虽然很多企业可能并没有意识到数据爆炸性增长带来问题的隐患，但是随着时间的推移，人们将越来越多地意识到数据对企业的重要性。大数据通常用来形容一个公司创造的大量非结构化和半结构化数据，这些数据在下载到关系数据库时用于分析会花费过多的时间和金钱。大数据分析常和云计算联系到一起，因为实时的大型数据分析需要像 MapReduce 一样的计算框架来向数十、数百甚至数千台计算机分配工作。

本书主要向读者介绍一种大规模数据处理的开源框架——Hadoop 生态系统。在深入探讨 Hadoop 的技术细节和应用之前，有必要花时间来了解 Hadoop 及其取得巨大成功的历史背景。Hadoop 并不是凭空想象出来的，它的出现源于人们创建和使用的数据量的爆炸性增长。在此背景下，不仅庞大的跨国公司面临着海量数据处理的困难，小型创业公司同样如此。与此同时，一些变革改变了软件和系统的部署方式，除了传统的基础设施，人们开始使用甚至偏好于分布式资源处理框架。

本书揭开了 Apache Hadoop 的神秘面纱，着重讲解了如何应用 Hadoop 和相关技术搭建工作系统并完成任务。本书共分为 9 章：第 1 章讲解 Hadoop 的生态系统，以及在行业中的应用场景；第 2 章讲解 Hadoop 分布式文件系统，包括 NameNode 和 DataNode 节点、机架感知策略、HDFS Shell 命令等；第 3 章讲解 MapReduce 并行计算框架，让读者了解 MapReduce 的工作原理；第 4 章讲解 HBase 分布式数据库，讲述了 HBase 如何实现数据存储、HBase 的节点类型、HBase API 开发；第 5 章讲解 Hive 数据仓库，介绍了 Hive 的架构、HQL 语法结构、Hive 数据查询案例；第 6 章讲解 Kafka 消息系统，介绍了 Kafka 消息系统的工作原理、Kafka 消息细节处理等；第 7 章讲解 Flume 日志处理系统，介绍了 Flume

的日志处理技术、Flume 如何进行流计算技术处理；第 8 章讲解 ZooKeeper 分布式协调系统，帮助读者理解如何实现 Hadoop 组件之间的协调控制；第 9 章讲解 Sqoop 数据迁移工具，涵盖了有效使用 Sqoop 处理实际场景中的数据迁移工作。

通过阅读本书，读者将迅速掌握编程概念，打下坚实的基础，并养成良好的习惯。此后，读者就可以开始了解其他大数据平台技术，如 Spark 内存计算框架、Flink 流批一体化处理平台，并能够更轻松地掌握大数据技术。

目 录

第 1 章　Hadoop 大数据平台概述 …… 1

1.1　Hadoop 大数据平台起源 …… 1
1.1.1　Hadoop 发展历程 …… 1
1.1.2　Hadoop 核心组件 …… 2
1.1.3　Hadoop 与云计算的关系 …… 3
1.2　Hadoop 集群搭建和简单应用 …… 3
1.2.1　集群服务器规划 …… 3
1.2.2　Hadoop 软件安装 …… 4
1.2.3　Hadoop 命令行的基本使用 …… 9
本章小结 …… 11

第 2 章　Hadoop 分布式文件系统 …… 12

2.1　HDFS 概述 …… 12
2.1.1　HDFS 的概念和特性 …… 12
2.1.2　HDFS 的局限性 …… 13
2.1.3　HDFS 保证可靠性的措施 …… 14
2.1.4　单点故障(单点失效)问题 …… 14
2.2　HDFS Shell 命令 …… 15
2.2.1　常见 Shell 命令 …… 15
2.2.2　其他 HDFS Shell 命令 …… 18
2.3　对 HDFS 的深入理解 …… 21
2.3.1　HDFS 的优点和缺点 …… 21
2.3.2　HDFS 的辅助功能 …… 22
2.4　HDFS 读写过程 …… 28
2.4.1　HDFS 写入数据过程 …… 28

2.4.2　HDFS 读取数据过程 … 29
2.5　分布式集群中 HDFS 的各种角色 … 30
  2.5.1　NameNode 的可靠性 … 30
  2.5.2　DataNode 的可靠性 … 31
  2.5.3　元数据的 CheckPoint … 31
本章小结 … 32

第 3 章　MapReduce 并行计算框架 … 33

3.1　MapReduce 概述 … 33
  3.1.1　为什么需要 MapReduce? … 33
  3.1.2　MapReduce 程序运行演示 … 34
  3.1.3　WordCount.java 源码分析 … 36
  3.1.4　编写自己的 WordCount 程序 … 39
3.2　MapReduce 的核心运行机制 … 43
3.3　MapReduce 的多 Job 串联和全局计数器 … 45
  3.3.1　MapReduce 的多 Job 串联 … 45
  3.3.2　全局计数器 … 46
  3.3.3　计数器该如何使用? … 50
  3.3.4　MapReduce 框架 Partitioner 分区 … 51
  3.3.5　MapReduce 框架 Combiner 分区 … 53
3.4　YARN 的资源调度 … 53
本章小结 … 56

第 4 章　HBase 分布式数据库 … 57

4.1　HBase 数据库概述 … 57
  4.1.1　HBase 数据库的使用场景 … 57
  4.1.2　HBase 数据库的安装 … 59
4.2　HBase 数据库物理架构 … 64
  4.2.1　HBase 集群节点类型 … 64
  4.2.2　HBase 数据存储 … 65
4.3　HBase 数据库操作 … 67
  4.3.1　HBase 命令行的启动 … 67
  4.3.2　HBase 表的操作 … 68

4.3.3　HBase 表中数据的操作 ……………………………………………… 71

　4.4　HBase 数据库的 API 操作 …………………………………………………… 73

　本章小结 …………………………………………………………………………… 83

第 5 章　Hive 数据仓库 …………………………………………………………… 84

　5.1　Hive 简介 ……………………………………………………………………… 84

　　5.1.1　什么是 Hive? ………………………………………………………… 84

　　5.1.2　Hive 的数据组织 ……………………………………………………… 86

　　5.1.3　Hive 的表类型 ………………………………………………………… 87

　5.2　Hive 的安装与使用 …………………………………………………………… 87

　　5.2.1　Hive 的安装配置 ……………………………………………………… 87

　　5.2.2　Hive 的基本使用 ……………………………………………………… 91

　　5.2.3　Hive 的连接方式 ……………………………………………………… 94

　5.3　Hive 数据结构 ………………………………………………………………… 96

　　5.3.1　Hive 数据类型 ………………………………………………………… 96

　　5.3.2　Hive 数据存储格式 …………………………………………………… 97

　　5.3.3　数据格式 ………………………………………………………………… 98

　5.4　Hive 数据操作 ………………………………………………………………… 98

　　5.4.1　管理库 …………………………………………………………………… 98

　　5.4.2　表操作 …………………………………………………………………… 101

　5.5　Hive 应用案例 ………………………………………………………………… 112

　　5.5.1　统计单月访问次数和总访问次数 …………………………………… 112

　　5.5.2　学生课程成绩统计 …………………………………………………… 116

　本章小结 …………………………………………………………………………… 130

第 6 章　Kafka 消息系统 ………………………………………………………… 132

　6.1　Kafka 消息系统的功能 ……………………………………………………… 132

　　6.1.1　Kafka 概述 ……………………………………………………………… 132

　　6.1.2　Kafka 组件架构 ………………………………………………………… 134

　　6.1.3　Kafka 软件安装 ………………………………………………………… 135

　　6.1.4　Kafka 服务的启动 ……………………………………………………… 137

　6.2　Kafka 组件术语 ……………………………………………………………… 138

　　6.2.1　主题与日志 ……………………………………………………………… 138

6.2.2　Kafka 日志处理 …………………………………………… 143
　　6.2.3　消息副本 ………………………………………………… 146
　　6.2.4　数据处理场景 …………………………………………… 149
　　6.2.5　生产者 …………………………………………………… 153
　　6.2.6　消费者 …………………………………………………… 155
本章小结 ………………………………………………………………… 158

第 7 章　Flume 日志处理系统 ……………………………………… 159

7.1　Flume 的简介 ……………………………………………………… 159
　　7.1.1　Flume 概述 ………………………………………………… 159
　　7.1.2　Flume NG 的介绍 ………………………………………… 160
　　7.1.3　Flume 的部署类型 ………………………………………… 161
7.2　Flume 的安装与配置 ……………………………………………… 164
　　7.2.1　Flume 的下载与安装 ……………………………………… 164
　　7.2.2　Flume Sources 描述 ……………………………………… 165
7.3　Flume 代理流配置 ………………………………………………… 167
　　7.3.1　单一代理流配置 …………………………………………… 167
　　7.3.2　单代理多流配置 …………………………………………… 167
　　7.3.3　配置多代理流程 …………………………………………… 167
　　7.3.4　多路复用流 ………………………………………………… 167
本章小结 ………………………………………………………………… 168

第 8 章　ZooKeeper 分布式协调系统 ……………………………… 169

8.1　分布式协调技术概述 ……………………………………………… 169
8.2　ZooKeeper 概述 …………………………………………………… 172
8.3　ZooKeeper 监听机制 ……………………………………………… 175
　　8.3.1　Watch 触发器 ……………………………………………… 175
　　8.3.2　监听原理 …………………………………………………… 176
　　8.3.3　ZooKeeper 应用举例 ……………………………………… 176
8.4　ZooKeeper 的安装与集群配置 …………………………………… 179
　　8.4.1　ZooKeeper 的安装 ………………………………………… 180
　　8.4.2　使用 ZooKeeper 命令的简单操作步骤 …………………… 186
本章小结 ………………………………………………………………… 188

第 9 章 Sqoop 数据迁移工具 190

9.1 Sqoop 功能概述 190
    9.1.1 Sqoop 软件介绍 190
    9.1.2 Sqoop 软件安装 191
9.2 Sqoop 命令操作 192
    9.2.1 Sqoop 的基本命令 192
    9.2.2 Sqoop 的数据导入 195
    9.2.3 将 MySQL 数据库中的表数据导入 Hive 199
    9.2.4 将 MySQL 数据库中的表数据导入 HBase 204
本章小结 204

参考文献 205

第 1 章
Hadoop 大数据平台概述

从大数据自身的技术体系来说,大数据所有的技术都紧紧围绕数据价值化来展开,企业对大数据的利用当前也逐渐从传统的数据采集和分析向数据生产转变,相信在工业互联网时代这一趋势会越发明显。

对于企业来说,借助于大数据来降低运营成本是一个重要的诉求,而通过大数据技术来降低运营成本的出发点非常多,不同行业企业要结合自身的实际情况来进行方案规划。当前很多企业利用大数据来构建自己的价值化考核体系,这是降耗提效的好方式。

大数据时代,数据的应用已经渗透到各行各业,但是传统的数据挖掘和分析已经不能满足行业发展的需求,大数据技术为企业业务分析和行业发展带来了新的思维角度,将会充分激发数据对社会发展的影响和推动。如何有效利用大数据平台?接下来我们就一起来了解 Apache Hadoop 大数据生态系统。

1.1 Hadoop 大数据平台起源

Hadoop 是 Hadoop 项目创建者 Doug Cutting 儿子的一只玩具的名字。他的儿子一直称呼一只黄色的大象玩具为 Hadoop,这刚好满足 Cutting 的命名需求——简短、容易拼写和发音、毫无意义、不会在别处被使用,于是 Hadoop 就诞生了。Hadoop 的发行版本有很多,有华为发行版、星环发行版、Intel 发行版、Cloudera 发行版(CDH)、MapR 版本以及 HortonWorks 版本等。所有发行版本都是基于 Apache Hadoop 衍生出来的,产生这些版本的原因可归结为 Apache Hadoop 的开源协议:任何人都可以对其进行修改,并作为开源或商业产品发布和销售。

1.1.1 Hadoop 发展历程

1. Hadoop 大数据平台的起源

① Hadoop 最早起源于 Nutch 项目,Nutch 的设计目标是构建一个大型的全网搜索引擎,包括网页抓取、索引、查询等功能,但随着抓取网页数量的增加,其遇到了严重的可扩展性问题——如何解决数十亿网页的存储和索引问题。

② 从 2003 年开始,Google 陆续发表的 3 篇论文为该问题提供了可行的解决方案。

- 分布式文件系统(DFS):可用于处理海量网页的存储问题。

- 分布式计算框架 MapReduce：可用于处理海量网页的索引计算问题。
- BigTable 分布式数据库：OLTP（联机事务处理，On-Line Transaction Processing）用于执行增、删、改操作，OLAP（联机分析处理，On-Line Analysis Processing）用于执行查询操作。

③ Nutch 的开发人员完成了相应的开源实现 HDFS 和 MapReduce，并将其从 Nutch 中剥离出来，成为独立项目 Hadoop。直到 2008 年 1 月，Hadoop 成为 Apache 顶级项目，迎来了快速发展期。

2. Hadoop 官网

我们可以通过 Hadoop 官网 http://hadoop.apache.org/来学习 Hadoop 的核心技术。Hadoop 大数据平台的处理主要就是存储和计算，我们安装 Hadoop 集群，目的是实现两个核心功能：一个操作系统 YARN 和一个分布式文件系统 HDFS，其实 MapReduce 就是运行在 YARN 之上的应用。

1.1.2　Hadoop 核心组件

Hadoop 是 Apache 旗下的一套开源软件平台，Hadoop 主要提供的功能是：利用服务器集群，根据用户自定义的逻辑对海量数据进行分布式处理。

1. Hadoop 的概念

① 狭义上：属于 Apache 基金会的一个顶级项目 Apache Hadoop。
② 广义上：以 Hadoop 为核心的整个大数据处理体系，包括计算和存储能力。

2. Hadoop 的核心组件

① Hadoop Common：支持其他 Hadoop 模块的常用工具。
② Hadoop 分布式文件系统（HDFS）：一种分布式文件系统，可提供对应用程序数据的高吞吐量访问。
③ Hadoop YARN：作业调度和集群资源管理的框架。
④ Hadoop MapReduce：一种用于并行处理大型数据集的基于 YARN 的系统。

3. Apache 的其他 Hadoop 相关项目

① Ambari：一种用于供应、管理和监控 Apache Hadoop 集群的基于 Web 的工具，其中包括对 HDFS、Hadoop MapReduce、Hive、HCatalog、HBase、ZooKeeper、Oozie、Pig 和 Sqoop 的支持。Ambari 还提供了一个用于查看集群运行状况的仪表板，如数据热图和可以直观地查看 MapReduce、Pig 和 Hive 应用程序的功能，以及以用户友好的方式诊断其性能特征的功能。

② Avro：数据序列化系统。
③ Cassandra：无单点故障的可扩展多主数据库。
④ Chukwa：管理大型分布式系统的数据收集系统。
⑤ HBase：可扩展的分布式数据库，支持大型表格的结构化数据存储。
⑥ Hive：提供数据汇总和即席查询的数据仓库基础架构。
⑦ Mahout：可扩展的机器学习和数据挖掘库。
⑧ Pig：用于并行计算的高级数据流语言和执行框架。
⑨ Spark：用于 Hadoop 数据的快速和通用计算引擎。Spark 提供了一个简单而富有表现力的编程模型，它支持广泛的应用程序，包括数据抽取、转换、加载（ETL），机器学习，流计算处

理和图计算。

⑩ Tez：一种基于 Hadoop YARN 的通用数据流编程框架，它提供了一个强大且灵活的引擎，可执行任意有向无环图（DAG）任务来处理批处理和交互式用例的数据。Hadoop、Pig 和 Hadoop 生态系统中的其他框架以及其他商业软件（如 ETL 工具）正在采用 Tez 来替代 Hadoop MapReduce 作为底层执行引擎。

⑪ ZooKeeper：分布式应用程序的高性能协调服务。

1.1.3　Hadoop 与云计算的关系

云计算是分布式计算、并行计算、网格计算、多核计算、网络存储、虚拟化、负载均衡等传统计算机技术和互联网技术融合发展的产物，其借助于基础设施即服务（IaaS）、平台即服务（PaaS）、软件即服务（SaaS）等业务模式，把强大的计算能力提供给终端用户。现阶段云计算的两大底层支撑技术为"虚拟化"和"大数据技术"，而 Hadoop 则是云计算的 PaaS 层的解决方案之一，并不等同于 PaaS，更不等同于云计算本身。

大数据与云计算密不可分。大数据必然无法用单台计算机进行处理，必须采用分布式计算架构。大数据的特色在于对海量数据的挖掘，但它必须依托云计算的分布式处理、分布式数据库、云存储和虚拟化技术。它们之间的关系可以这样来理解：云计算技术就是一个容器，大数据正是存放在这个容器中的水，大数据是要依靠云计算技术来进行存储和计算的。大数据发展具有如下趋势。

趋势一：数据的资源化。

资源化是指大数据成为企业和社会关注的重要战略资源，并已成为大家争相抢夺的新焦点。因而，企业必须要提前制订大数据营销战略计划，抢占市场先机。

趋势二：与云计算的深度结合。

大数据离不开云计算技术，云计算为大数据提供了弹性可拓展的基础设备，是产生大数据的平台之一。自 2013 年起，大数据技术已开始和云计算技术紧密结合，未来两者的关系将更为密切。除此之外，物联网、移动互联网等新兴计算形态也将一齐助力大数据革命，让大数据营销发挥出更大的影响力。

趋势三：科学理论的突破。

随着大数据的快速发展，就像计算机和互联网一样，大数据很有可能是新一轮的技术革命。随之兴起的数据挖掘、机器学习和人工智能等相关技术，可能会改变数据世界里的很多算法和基础理论，实现科学技术上的突破。

1.2　Hadoop 集群搭建和简单应用

1.2.1　集群服务器规划

1. 节点规划

本教材中，读者可以使用 4 台 CentOS 6.7 虚拟机进行集群搭建。为了方便呈现每台主机的功能,主机角色和 IP 地址设置等参考表 1-1。

表 1-1 节点规划

服务器	IP	用户	HDFS	YARN
hadoop1	192.168.123.102	hadoop	NameNode,DataNode	NodeManager
hadoop2	192.168.123.103	hadoop	DataNode	NodeManager
hadoop3	192.168.123.104	hadoop	DataNode,SecondaryNameNode	NodeManager
hadoop4	192.168.123.105	hadoop	DataNode	ResourceManager,NodeManager

2. Hadoop 集群的部署模式

Hadoop 的运行模式分为 3 种:本地运行模式、伪分布运行模式、集群运行模式。

(1) 独立模式(即本地运行模式)

无须运行任何守护进程,所有程序都在单个 Java 虚拟机(JVM)上执行。由于在本机模式下测试和调试 MapReduce 程序较为方便,因此这种模式适合用在开发阶段。独立模式无须配置任何文件。

(2) 伪分布运行模式

如果 Hadoop 对应的 Java 进程都运行在一个物理机器上,则称为伪分布运行模式。以 Windows 为例,在其他系统下,需要修改路径。

(3) 集群运行模式

集群中每一个节点都可以独立运行 Hadoop 的相关进程,防止单点故障,适合应用于生产环境。

1.2.2 Hadoop 软件安装

1. 实验环境规划
- 规划安装用户:hadoop 用户。
- 规划安装目录:/home/hadoop/apps。
- 规划数据目录:/home/hadoop/data。

注意:apps 和 data 两个目录需要自己单独创建。

2. 上传安装软件,并实现软件解压缩

使用 hadoop 用户,尽量不使用 CentOS 操作系统 root 用户登录。

```
[hadoop@hadoop1 apps]$ ls
hadoop-2.7.5-centos-6.7.tar.gz
[hadoop@hadoop1 apps]$ tar -zxvf hadoop-2.7.5-centos-6.7.tar.gz
```

3. 修改配置文件

配置文件目录:/home/hadoop/apps/hadoop-2.7.5/etc/hadoop。

(1) hadoop-env.sh 配置文件

```
[hadoop@hadoop1 hadoop]$ vi hadoop-env.sh
```

Hadoop 使用 Java 开发环境,需要修改操作系统 JAVA_HOME 环境变量:

```
export JAVA_HOME = /usr/local/jdk1.8.0_73
```

(2) core-site.xml 配置文件

```
[hadoop@hadoop1 hadoop]$ vi core-site.xml
```

- fs.defaultFS：这个属性用来指定 NameNode 的 HDFS 协议的文件系统通信地址，可以指定为一个主机＋端口，也可以指定为一个 NameNode 服务〔这个服务内部可以有多台 NameNode 实现高可用性双机集群（HA）的 NameNode 服务〕。
- hadoop.tmp.dir：Hadoop 集群在工作时存储的一些临时文件的目录。

参考配置如下：

```xml
<configuration>
    <property>
        <name>fs.defaultFS</name>
        <value>hdfs://hadoop1:9000</value>
    </property>
    <property>
        <name>hadoop.tmp.dir</name>
        <value>/home/hadoop/data/hadoopdata</value>
    </property>
</configuration>
```

(3) hdfs-site.xml 配置文件

```
[hadoop@hadoop1 hadoop]$ vi hdfs-site.xml
```

- dfs.namenode.name.dir：NameNode 元数据的存放目录，记录了 HDFS 中文件的元数据。
- dfs.datanode.data.dir：DataNode 数据的存放目录，也就是数据块（Block）的存放目录。
- dfs.replication：HDFS 的副本数设置。文件写入时被分割为 Block 后，每个 Block 的冗余副本个数，默认配置是 3。
- dfs.secondary.http.address：SecondaryNameNode 运行节点的信息，可以和 NameNode 不同节点，也可以和 NameNode 同一节点。

参考配置如下：

```xml
<configuration>
    <property>
        <name>dfs.namenode.name.dir</name>
        <value>/home/hadoop/data/hadoopdata/name</value>
        <description>可以配置多个不同目录</description>
    </property>
    <property>
        <name>dfs.datanode.data.dir</name>
        <value>/home/hadoop/data/hadoopdata/data</value>
        <description>DataNode 的数据存储目录</description>
    </property>
    <property>
```

```xml
        <name>dfs.replication</name>
        <value>2</value>
        <description>HDFS 数据块的副本存储个数,默认是 3</description>
    </property>
    <property>
        <name>dfs.secondary.http.address</name>
        <value>hadoop3:50090</value>
        <description>SecondaryNameNode 运行节点的信息,可以和 NameNode 不同节点,也可以和 NameNode 相同节点</description>
    </property>
</configuration>
```

(4) mapred-site.xml 配置文件

```
[hadoop@hadoop1 hadoop]$ cp mapred-site.xml.template mapred-site.xml
[hadoop@hadoop1 hadoop]$ vi mapred-site.xml
```

MapReduce.framework.name:指定 MapReduce 框架为 YARN 方式,Hadoop 二代 MP 也基于资源管理系统 YARN 来运行。

参考配置如下:

```xml
<configuration>
    <property>
        <name>MapReduce.framework.name</name>
        <value>yarn</value>
    </property>
</configuration>
```

(5) yarn-site.xml 配置文件

```
[hadoop@hadoop1 hadoop]$ vi yarn-site.xml
```

- yarn.resourcemanager.hostname:YARN 总管理器的进程间通信(IPC)地址。
- yarn.nodemanager.aux-services:NodeManager 上的附属服务,需配置成 MapReduce_shuffle,才可运行 MapReduce 程序。

参考配置如下:

```xml
<configuration>
<!-- Site specific YARN configuration properties -->
    <property>
        <name>yarn.resourcemanager.hostname</name>
        <value>hadoop4</value>
    </property>
    <property>
        <name>yarn.nodemanager.aux-services</name>
        <value>MapReduce_shuffle</value>
        <description>shuffle service</description>
    </property>
</configuration>
```

（6）slaves 配置文件

```
[hadoop@hadoop1 hadoop]$ vi slaves
```

参考配置如下：

```
hadoop1
hadoop2
hadoop3
hadoop4
```

4. 把安装包分发给其他的节点

每台服务器中的 Hadoop 安装包的目录必须一致，安装包的配置信息还必须保持一致：

```
[hadoop@hadoop1 hadoop]$ scp -r ~/apps/hadoop-2.7.5/ hadoop2:~/apps/
[hadoop@hadoop1 hadoop]$ scp -r ~/apps/hadoop-2.7.5/ hadoop3:~/apps/
[hadoop@hadoop1 hadoop]$ scp -r ~/apps/hadoop-2.7.5/ hadoop4:~/apps/
```

5. 配置 Hadoop 环境变量

因为我们使用 hadoop 用户进行安装，所以编辑个人目录下的 .bashrc 文件，如下：

```
[hadoop@hadoop1 ~]$ vi .bashrc
export HADOOP_HOME=/home/hadoop/apps/hadoop-2.7.5
export PATH=$PATH:$HADOOP_HOME/bin:$HADOOP_HOME/sbin:
```

使环境变量生效：

```
[hadoop@hadoop1 bin]$ source ~/.bashrc
```

6. 查看 Hadoop 版本

命令及输出内容参考如下：

```
[hadoop@hadoop1 bin]$ hadoop version
hadoop 2.7.5
Subversion Unknown -r Unknown
Compiled by root on 2020-3-2T05:30Z
Compiled with protoc 2.5.0
From source with checksum 9f118f95f47043332d51891e37f736e9
This command was run using /home/hadoop/apps/hadoop-2.7.5/share/hadoop/common/hadoop-common-2.7.5.jar
```

7. Hadoop 集群初始化

HDFS 初始化只能在主节点上进行，执行过程中会产生大量控制台输出提示，命令及最后输出的内容参考如下：

```
[hadoop@hadoop1 ~]$ hadoop namenode -format
DEPRECATED: Use of this script to execute hdfs command is deprecated.
Instead use the hdfs command for it.
20/03/03 11:13:24 INFO namenode.namenode: STARTUP_MSG:
/************************************************************
  20/03/03 11:13:26 INFO namenode.FSImageFormatProtobuf: Image file /home/hadoop/data/hadoopdata/name/current/fsimage.ckpt_0000000000000000000 of size 323 bytes saved in 0 seconds.
  20/03/03 11:13:26 INFO namenode.NNStorageRetentionManager: Going to retain 1 images with txid >= 0
```

```
20/03/03 11:13:26 INFO util.ExitUtil: Exiting with status 0
20/03/03 11:13:26 INFO namenode.namenode: SHUTDOWN_MSG:
/************************************************************
SHUTDOWN_MSG: Shutting down namenode at hadoop1/192.168.123.102
************************************************************/
```

8. 启动 HDFS 相关服务和进程

① 启动 HDFS 分布式文件系统服务,值得注意的是:在 Hadoop 集群的任一节点上都可以启动。命令及输出内容参考如下:

```
[hadoop@hadoop1 ~]$ start-dfs.sh
Starting namenodes on [hadoop1]
hadoop1: starting namenode, logging to /home/hadoop/apps/hadoop-2.7.5/logs/hadoop-hadoop-namenode-hadoop1.out
hadoop3: starting datanode, logging to /home/hadoop/apps/hadoop-2.7.5/logs/hadoop-hadoop-datanode-hadoop3.out
hadoop2: starting datanode, logging to /home/hadoop/apps/hadoop-2.7.5/logs/hadoop-hadoop-datanode-hadoop2.out
hadoop4: starting datanode, logging to /home/hadoop/apps/hadoop-2.7.5/logs/hadoop-hadoop-datanode-hadoop4.out
hadoop1: starting datanode, logging to /home/hadoop/apps/hadoop-2.7.5/logs/hadoop-hadoop-datanode-hadoop1.out
Starting secondary namenodes [hadoop3]
hadoop3: starting secondarynamenode, logging to /home/hadoop/apps/hadoop-2.7.5/logs/hadoop-hadoop-secondarynamenode-hadoop3.out
```

② 启动 YARN 资源管理器,值得注意的是:只能在主节点上进行启动。命令及输出内容参考如下:

```
[hadoop@hadoop4 ~]$ start-yarn.sh
starting yarn daemons
starting resourcemanager, logging to /home/hadoop/apps/hadoop-2.7.5/logs/yarn-hadoop-resourcemanager-hadoop4.out
hadoop2: starting nodemanager, logging to /home/hadoop/apps/hadoop-2.7.5/logs/yarn-hadoop-nodemanager-hadoop2.out
hadoop3: starting nodemanager, logging to /home/hadoop/apps/hadoop-2.7.5/logs/yarn-hadoop-nodemanager-hadoop3.out
hadoop4: starting nodemanager, logging to /home/hadoop/apps/hadoop-2.7.5/logs/yarn-hadoop-nodemanager-hadoop4.out
hadoop1: starting nodemanager, logging to /home/hadoop/apps/hadoop-2.7.5/logs/yarn-hadoop-nodemanager-hadoop1.out
```

9. 查看每台服务器的进程

hadoop1 使用 jps 命令查看服务器进程。命令及输出内容参考如下:

```
[hadoop@hadoop1 ~]$ jps
3680 Jps
3557 NodeManager
```

```
2775 NameNode
2908 DataNode
```

10. 启动 HDFS 和 YARN 的 Web 管理界面

读者可以通过"http://192.168.123.102:50070"来访问 HDFS 集群服务，也可以通过访问数据节点"http://数据节点或者节点 IP:8088"来查看 YARN 资源调度服务。

① HDFS 集群界面如图 1-1 所示。

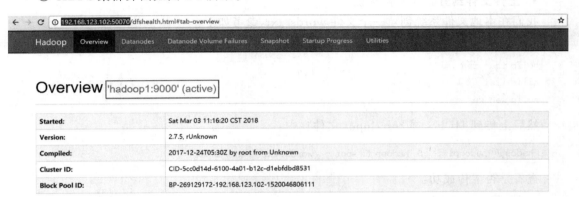

图 1-1　HDFS 集群界面

② YARN 管理界面如图 1-2 所示。

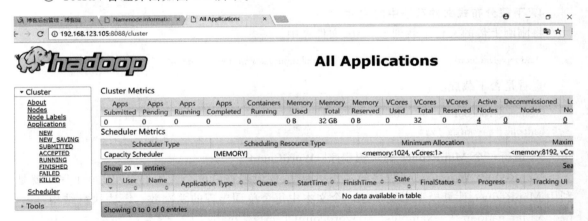

图 1-2　YARN 管理界面

1.2.3　Hadoop 命令行的基本使用

1. 创建文件夹

在 HDFS 上创建一个文件夹/test/input：

[hadoop@hadoop1 ~]$ hadoop fs -mkdir -p /test/input

2. 查看创建的文件夹

[hadoop@hadoop1 ~]$ hadoop fs -ls /

```
Found 1 items
drwxr-xr-x   -hadoop supergroup          0 2020-03-03 11:33 /test
[hadoop@hadoop1 ~]$ hadoop fs -ls /test
Found 1 items
drwxr-xr-x   -hadoop supergroup          0 2020-03-03 11:33 /test/input
[hadoop@hadoop1 ~]$
```

3. 上传文件到分布式文件系统

我们在本地文件系统创建一个文件 words.txt,参考如下:

```
[hadoop@hadoop1 ~]$ vi words.txt
hello zhangsan
hello lisi
hello wangwu
```

然后上传到 HDFS 的 /test/input 文件夹中:

```
[hadoop@hadoop1 ~]$ hadoop fs -put ~/words.txt /test/input
```

查看是否上传成功:

```
[hadoop@hadoop1 ~]$ hadoop fs -ls /test/input
Found 1 items
-rw-r--r--   2 hadoop supergroup         39 2020-03-03 11:37 /test/input/words.txt
```

4. 下载分布式文件系统中的文件

将刚刚上传的文件下载到 ~/data 文件夹中:

```
[hadoop@hadoop1 ~]$ hadoop fs -get /test/input/words.txt ~/data
```

查看是否下载成功:

```
[hadoop@hadoop1 ~]$ ls data
hadoopdata  words.txt
```

5. 运行一个 MapReduce 的示例程序:WordCount

MapReduce WordCount 程序的主要功能是:假设现在有 n 个文本,WordCount 程序就是利用 MapReduce 计算模型来统计这 n 个文本中每个单词出现的总次数。语法执行结构:bin/hadoop jar hadoop-*-examples.jar wordcount [-m <# maps>] [-r <# reducers>] <in-dir> <out-dir>。

我们执行如下命令,进行 /test/input 目录的文件词频统计:

```
[hadoop@hadoop1 ~]$ hadoop jar ~/apps/hadoop-2.7.5/share/hadoop/MapReduce/hadoop-MapReduce-examples-2.7.5.jar wordcount /test/input /test/output
```

在 YARN Web 界面可以查看执行的进程,如图 1-3 所示。

接下来,查看词频统计结果,我们可以查看到每个单词出现的次数:

```
[hadoop@hadoop1 ~]$ hadoop fs -ls /test/output
Found 2 items
-rw-r--r--   2 hadoop supergroup          0 2020-03-03 11:42 /test/output/_SUCCESS
-rw-r--r--   2 hadoop supergroup         35 2020-03-03 11:42 /test/output/part-r-00000
```

第 1 章　Hadoop 大数据平台概述

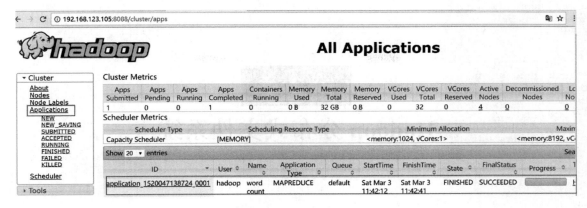

图 1-3　查看执行的进程

```
[hadoop@hadoop1 ~]$ hadoop fs -cat /test/output/part-r-00000
hello      3
lisi       1
wangwu     1
zhangsan   1
```

本 章 小 结

① Hadoop 在可伸缩性、健壮性、计算性能和成本上具有无可替代的优势，事实上其已成为当前互联网企业主流的大数据分析平台。Hadoop 是 Apache 软件基金会旗下的一个开源分布式计算平台。以 Hadoop 分布式文件系统和 MapReduce（Google MapReduce 的开源实现）为核心的 Hadoop 为用户提供了系统底层细节透明的分布式基础架构。

② Hadoop 集群安装配置模式结合不同的场景大致可以分为以下 3 种情况。

- 单机模式：表示所有的分布式系统都是单机的。
- 伪分布式模式：表示集群中的所有角色都分配给了一个节点，整个集群被安装在只有一个节点的集群中。主要用于做快速使用或者测试的效果，模拟多个服务进程环境。
- 分布式模式：表示集群中的节点会被分配成很多种角色，分散在整个集群中。主要用于企业项目实战等场景中。

③ HDFS 和 MapReduce 是 Hadoop 的两个重要核心，其中 MapReduce 是 Hadoop 的分布式计算模型，MapReduce 主要分为两步：Map 步和 Reduce 步。

第 2 章

Hadoop 分布式文件系统

2.1 HDFS 概述

2.1.1 HDFS 的概念和特性

1. HDFS 的概念

HDFS(Hadoop Distributed File System)是 Hadoop 分布式文件系统,主要用于解决海量数据的存储问题。首先,它是一个文件系统,用于存储文件,通过统一的命名空间——目录树来定位文件;其次,它具备分布式存储的特点,由很多服务器联合起来实现其功能,集群中的服务器有各自的角色。

2. HDFS 的重要特性

① HDFS 中的文件在物理上是分块存储,块的大小可以通过配置参数 dfs.blocksize 来规定,默认大小在 Hadoop 1.x 版本中是 64 MB,在 Hadoop 2.x 版本中是 128 MB,在 Hadoop 3.x 版本中是 256 MB,HDFS 不太适合小文件的存储场景。

② HDFS 会给客户端提供一个统一的抽象目录树,客户端通过路径来访问文件,形如:hdfs://namenode:port/dir-a/dir-b/dir-c/file.data。

③ HDFS 目录结构及文件分块信息(元数据)的管理由 NameNode 节点承担。NameNode 是 HDFS 集群的主(Master)节点,负责维护整个 HDFS 的目录树,以及每一个路径(文件)所对应的 Block 信息(Block 的 ID 及所在的 DataNode 服务器)。

④ 文件的各个 Block 的存储管理由 DataNode 节点承担。DataNode 是 HDFS 集群的从(Slave)节点,每一个 Block 都可以在多个 DataNode 上存储多个副本,副本数量可以通过参数 dfs.replication 设置。

⑤ HDFS 是设计成适应一次写入、多次读出的场景,且不支持文件的修改。

3. 图解 HDFS

通过上面的描述我们知道,HDFS 具有以下特点。

① 保存多个副本,且提供容错机制,副本丢失或宕机自动恢复(默认存储 3 份)。

② Hadooop 采用 Linux 作为底层操作系统,可以运行在廉价的机器上,具有良好的兼容性。

③ 适合大数据的处理。HDFS 默认会将文件分割成 Block(在 Hadoop 2.x 版本中默认 128 MB 为 1 个 Block),然后将 Block 按键值对存储在 HDFS 上,并将键值对的映射存到内存中。如果小文件太多,则内存的负担会很重。

如图 2-1 所示,HDFS 是按照 Master 和 Slave 的结构,分为 NameNode、SecondaryNameNode、DataNode 这几个角色。

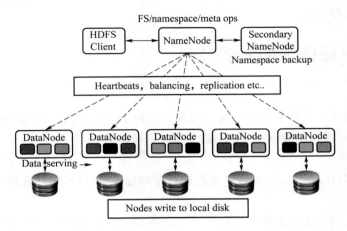

图 2-1　HDFS 架构

- NameNode:是 Master 节点,负责管理数据块映射、处理客户端的读写请求、配置副本策略、管理 HDFS 的名称空间。
- SecondaryNameNode:是一个加速主节点启动的服务器,分担 NameNode 的工作量,是 NameNode 的冷备份,目的是合并 fsimage 和 edits 然后再发送给 NameNode。
- DataNode:是 Slave 节点,负责存储 Client 发来的数据块,执行数据块的读写操作。
- fsimage:元数据镜像文件(文件系统的目录树)。
- edits:元数据的操作日志(针对文件系统做的修改操作记录)。

NameNode 内存中存储的是元数据镜像文件 fsimage 和元数据的操作日志 edits。SecondaryNameNode 负责定时(默认 1 小时)从 NameNode 上获取 fsimage 和 edits 进行合并,然后再发送给 NameNode,减少 NameNode 的工作量。

2.1.2　HDFS 的局限性

1. 无法提供低延时数据访问

在用户交互性的应用中,需要在毫秒级或者几秒的时间内得到响应。由于 HDFS 为高吞吐率做了设计,因此牺牲了快速响应。对于低延时的应用,可以考虑使用 HBase 或者 Cassandra。

2. 不适合大量的小文件存储

标准的 HDFS 数据块的大小是 128 MB,存储小文件并不会浪费实际的存储空间,但是无疑会增加在 NameNode 节点上的元数据,大量的小文件会影响整个集群的性能。相对来说,Linux Btrfs 为小文件做了优化——inline file,对于小文件有很好的空间优化和访问时间优化作用。

3. 不支持多用户写入、修改文件

HDFS 的文件只能有一个写入者，而且写操作只能在文件结尾以追加的方式进行。HDFS 不支持多个写入者，也不支持在写入后对文件任意位置进行修改。

但是大数据领域分析的是已经存在的数据，这些数据一旦产生就不需要修改，因此 HDFS 的这些特性和设计局限就很容易理解了。HDFS 为大数据领域的数据分析提供了非常重要且十分基础的文件存储功能。

2.1.3　HDFS 保证可靠性的措施

1. 冗余备份

每个文件存储成一系列数据块。为了实现容错功能，文件的所有数据块都会有副本（副本数量即复制因子，可通过调整 dfs.replication 参数进行配置）。

2. 副本存放

采用机架感知（rack-aware）的策略来改进数据的可靠性、高可用性和网络带宽的利用率。

3. 心跳检测

NameNode 周期性地从集群中的每一个 DataNode 接收心跳包和块报告，如果收到了心跳包，则说明该 DataNode 工作正常。

4. 安全模式

系统启动时，NameNode 会进入安全模式，此时不会出现数据块的写操作。

5. 数据完整性检测

HDFS 客户端软件实现了对 HDFS 文件内容的校验和 Checksum 检查 dfs.bytes-per-checksum。

2.1.4　单点故障（单点失效）问题

1. 单点故障问题

一个 HDFS 集群一般包括一个 NameNode 节点和多个 DataNode 节点。NameNode 节点是管理文件命名空间和调节客户端访问文件的主服务器，DataNode 节点是被 NameNode 节点控制并存储实际数据的从服务器。作为 HDFS 集群的关键节点，NameNode 节点的单点故障成为该分布式系统面临的最大风险。如果 NameNode 节点失效，那么客户端或 MapReduce 作业均无法读写、查看文件。

2. 解决方案

① 启动一个拥有文件系统元数据的新 NameNode 节点（这个方案一般不采用，因为复制元数据非常耗时间）。

② 配置一对活动-备用（active-standby）NameNode 节点，活动 NameNode 节点失效时，备用 NameNode 节点立即接管，用户不会有明显的中断感觉。

- 共享编辑日志文件〔借助于网络文件系统（NFS）、ZooKeeper 协调机制等〕。
- DataNode 同时向两个 NameNode 汇报数据块信息。
- 客户端采用特定机制处理 NameNode 失效问题，该机制对用户透明。

2.2 HDFS Shell 命令

2.2.1 常见 Shell 命令

1. 查看 HDFS 根目录

[hadoop@hadoop1 ~]$ hadoop fs -ls /
Found 2 items
drwxr-xr-x   -hadoop supergroup   0 2020-03-03 11:42 /test
drwx------   -hadoop supergroup   0 2020-03-03 11:42 /tmp

2. 创建文件夹并查看

[hadoop@hadoop1 ~]$ hadoop fs -mkdir /test1
[hadoop@hadoop1 ~]$ hadoop fs -ls /
Found 3 items
drwxr-xr-x   -hadoop supergroup   0 2020-03-08 11:09 /test1
drwxr-xr-x   -hadoop supergroup   0 2020-03-03 11:42 /test
drwx------   -hadoop supergroup   0 2020-03-03 11:42 /tmp

3. 级联创建文件夹

[hadoop@hadoop1 ~]$ hadoop fs -mkdir -p /aa/bb/cc

查看 HDFS 根目录下的所有文件夹内容，包括子文件夹里面的文件夹：

[hadoop@hadoop1 ~]$ hadoop fs -ls -R /aa
drwxr-xr-x   -hadoop supergroup   0 2020-03-08 11:12 /aa/bb
drwxr-xr-x   -hadoop supergroup   0 2020-03-08 11:12 /aa/bb/cc

4. 上传文件

[hadoop@hadoop1 ~]$ ls
apps   data   words.txt
[hadoop@hadoop1 ~]$ hadoop fs -put words.txt /aa
[hadoop@hadoop1 ~]$ hadoop fs -copyFromLocal words.txt /aa/bb
[hadoop@hadoop1 ~]$

5. 下载文件

[hadoop@hadoop1 ~]$ hadoop fs -get /aa/words.txt ~/newwords.txt
[hadoop@hadoop1 ~]$ ls
apps   data   newwords.txt   words.txt
[hadoop@hadoop1 ~]$ hadoop fs -copyToLocal /aa/words.txt ~/newwords1.txt
[hadoop@hadoop1 ~]$ ls
apps   data   newwords1.txt   newwords.txt   words.txt
[hadoop@hadoop1 ~]$

6. 合并下载

[hadoop@hadoop1 ~]$ hadoop fs -getmerge /aa/words.txt /aa/bb/words.txt ~/2words.txt

```
[hadoop@hadoop1 ~]$ ll
总用量 24
-rw-r--r--.  1 hadoop hadoop   78 3月  8 12:42 2words.txt
drwxrwxr-x.  3 hadoop hadoop 4096 3月  3 10:30 apps
drwxrwxr-x.  3 hadoop hadoop 4096 3月  3 11:40 data
-rw-r--r--.  1 hadoop hadoop   39 3月  8 11:49 newwords1.txt
-rw-r--r--.  1 hadoop hadoop   39 3月  8 11:48 newwords.txt
-rw-rw-r--.  1 hadoop hadoop   39 3月  3 11:31 words.txt
[hadoop@hadoop1 ~]$
```

7. 复制文件

复制文件过程是从 HDFS 的一个路径复制到 HDFS 的另一个路径:

```
[hadoop@hadoop1 ~]$ hadoop fs -ls /test1
[hadoop@hadoop1 ~]$ hadoop fs -cp /aa/words.txt /test1
[hadoop@hadoop1 ~]$ hadoop fs -ls /test1
```

```
Found 1 items
-rw-r--r--   2 hadoop supergroup         39 2020-03-08 12:46 /test1/words.txt
```

8. 移动文件

```
[hadoop@hadoop1 ~]$ hadoop fs -ls /aa/bb/cc
[hadoop@hadoop1 ~]$ hadoop fs -mv /test1/words.txt /aa/bb/cc
[hadoop@hadoop1 ~]$ hadoop fs -ls /aa/bb/cc
```

```
Found 1 items
-rw-r--r--   2 hadoop supergroup         39 2020-03-08 12:46 /aa/bb/cc/words.txt
```

9. 删除文件

删除文件或文件夹:

```
[hadoop@hadoop1 ~]$ hadoop fs -rm /aa/bb/cc/words.txt
20/03/08 12:49:08 INFO fs.TrashPolicyDefault: NameNode trash configuration: Deletion interval = 0 minutes, Emptier interval = 0 minutes.
Deleted /aa/bb/cc/words.txt
```

删除空目录:

```
[hadoop@hadoop1 ~]$ hadoop fs -rmdir /aa/bb/cc/
[hadoop@hadoop1 ~]$ hadoop fs -ls /aa/bb/
```

```
Found 1 items
-rw-r--r--   2 hadoop supergroup         39 2020-03-08 11:43 /aa/bb/words.txt
```

强制删除:

```
[hadoop@hadoop1 ~]$ hadoop fs -rm /aa/bb/
rm: '/aa/bb': Is a directory
[hadoop@hadoop1 ~]$ hadoop fs -rm -r /aa/bb/
20/03/08 12:51:31 INFO fs.TrashPolicyDefault: NameNode trash configuration: Deletion interval = 0 minutes, Emptier interval = 0 minutes.
Deleted /aa/bb
```

```
[hadoop@hadoop1 ~]$ hadoop fs -ls /aa
Found 1 items
-rw-r--r--   2 hadoop supergroup         39 2020-03-08 11:41 /aa/words.txt
```

10. 从本地剪切文件到 HDFS 上

```
[hadoop@hadoop1 ~]$ ls
```

```
apps    data    hello.txt
```

```
[hadoop@hadoop1 ~]$ hadoop fs -moveFromLocal ~/hello.txt /aa
```

```
[hadoop@hadoop1 ~]$ ls
```

```
apps    data
```

11. 追加文件

追加 hello.txt 内容到 words.txt 之前，words.txt 文件大小如图 2-2 所示。

图 2-2 追加内容前的文件大小

```
[hadoop@hadoop1 ~]$ hadoop fs -appendToFile ~/hello.txt /aa/words.txt
```

追加 hello.txt 内容到 words.txt 之后，words.txt 文件大小如图 2-3 所示。

12. 查看文件内容

```
[hadoop@hadoop1 ~]$ hadoop fs -cat /aa/hello.txt
```

```
hello

hello

hello
```

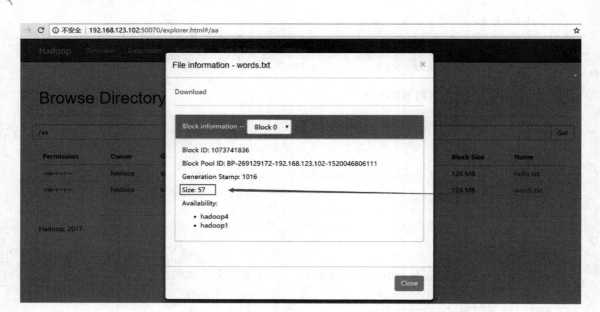

图 2-3　追加内容后的文件大小

2.2.2　其他 HDFS Shell 命令

1. chgrp

使用方法：hadoop fs -chgrp [-R] GROUP URI [URI …]。

功能：改变文件所属的组。使用-R 将使改变在目录结构下递归进行,命令的使用者必须是文件的所有者或者超级用户。更多的信息请参见 HDFS 权限用户指南。

2. chmod

使用方法：hadoop fs -chmod [-R] < MODE[,MODE]… | OCTALMODE > URI [URI …]。

功能：改变文件的权限。使用-R 将使改变在目录结构下递归进行,命令的使用者必须是文件的所有者或者超级用户。

3. chown

使用方法：hadoop fs -chown [-R] [OWNER][:[GROUP]] URI [URI]。

功能：改变文件的拥有者。使用-R 将使改变在目录结构下递归进行,命令的使用者必须是超级用户。

4. du

使用方法：hadoop fs -du URI [URI …]。

功能：显示分布式文件系统目录中所有文件的大小,或者当只指定一个文件时,显示此文件的大小。

示例：

hadoop fs -du /user/hadoop/dir1 /user/hadoop/file1 hdfs://host:port/user/hadoop/dir1

返回值：成功返回 0,失败返回 −1。

5. dus

使用方法：hadoop fs -dus < args >。

功能：显示文件的大小。

6. expunge

使用方法：hadoop fs -expunge。

功能：清空回收站。

7. setrep

使用方法：hadoop fs -setrep [-R] < path >。

功能：改变一个文件的副本系数。-R 选项用于递归改变目录下所有文件的副本系数。

示例：

```
hadoop fs -setrep -w 3 -R /user/hadoop/dir1
```

返回值：成功返回 0，失败返回 —1。

8. tail

使用方法：hadoop fs -tail [-f] URI。

功能：将文件尾部 1 KB 的内容输出到 stdout，支持-f 选项，行为和 UNIX 中一致。

示例：

```
hadoop fs -tail pathname
```

返回值：成功返回 0，失败返回 —1。

9. test

使用方法：hadoop fs -test -[ezd] URI。

选项：

-e：检查文件是否存在。如果存在则返回 0。

-z：检查文件是否是 0 字节。如果是则返回 0。

-d：如果路径是一个目录，则返回 1，否则返回 0。

示例：

```
hadoop fs -test -e filename
```

查看集群的工作状态：

```
[hadoop@hadoop1 ~]$ hdfs dfsadmin -report
Configured Capacity: 73741402112 (68.68 GB)
Present Capacity: 52781039616 (49.16 GB)
DFS Remaining: 52780457984 (49.16 GB)
DFS Used: 581632 (568 KB)
DFS Used%: 0.00%
Under replicated blocks: 0
Blocks with corrupt replicas: 0
Missing blocks: 0
Missing blocks (with replication factor 1): 0
-------------------------------------------------
Live datanodes (4):
```

Name：192.168.123.102:50010 (Hadoop1)
Hostname:hadoop1
Decommission Status : Normal
Configured Capacity: 18435350528 (17.17 GB)
DFS Used: 114688 (112 KB)
Non DFS Used: 4298661888 (4.00 GB)
DFS Remaining: 13193277440 (12.29 GB)
DFS Used%: 0.00%
DFS Remaining%: 71.57%
Configured Cache Capacity: 0 (0 B)
Cache Used: 0 (0 B)
Cache Remaining: 0 (0 B)
Cache Used%: 100.00%
Cache Remaining%: 0.00%
Xceivers: 1
Last contact: Thu Mar 08 13:05:11 CST 2020
Name：192.168.123.105:50010 (Hadoop4)
Hostname:hadoop4
Decommission Status : Normal
Configured Capacity: 18435350528 (17.17 GB)
DFS Used: 49152 (48 KB)
Non DFS Used: 4295872512 (4.00 GB)
DFS Remaining: 13196132352 (12.29 GB)
DFS Used%: 0.00%
DFS Remaining%: 71.58%
Configured Cache Capacity: 0 (0 B)
Cache Used: 0 (0 B)
Cache Remaining: 0 (0 B)
Cache Used%: 100.00%
Cache Remaining%: 0.00%
Xceivers: 1
Last contact: Thu Mar 08 13:05:13 CST 2020
Name：192.168.123.103:50010 (Hadoop2)
Hostname:hadoop2
Decommission Status : Normal
Configured Capacity: 18435350528 (17.17 GB)
DFS Used: 233472 (228 KB)
Non DFS Used: 4295700480 (4.00 GB)
DFS Remaining: 13196120064 (12.29 GB)
DFS Used%: 0.00%
DFS Remaining%: 71.58%
Configured Cache Capacity: 0 (0 B)
Cache Used: 0 (0 B)
Cache Remaining: 0 (0 B)
Cache Used%: 100.00%
Cache Remaining%: 0.00%

```
Xceivers: 1
Last contact: Thu Mar 08 13:05:11 CST 2020
Name: 192.168.123.104:50010（Hadoop3）
Hostname:hadoop3
Decommission Status : Normal
Configured Capacity: 18435350528 (17.17 GB)
DFS Used: 184320（180 KB）
Non DFS Used: 4296941568（4.00 GB）
DFS Remaining: 13194928128（12.29 GB）
DFS Used％: 0.00％
DFS Remaining％: 71.57％
Configured Cache Capacity: 0（0 B）
Cache Used: 0（0 B）
Cache Remaining: 0（0 B）
Cache Used％: 100.00％
Cache Remaining％: 0.00％
Xceivers: 1
Last contact: Thu Mar 08 13:05:10 CST 2020
```

2.3 对 HDFS 的深入理解

2.3.1 HDFS 的优点和缺点

1. HDFS 的优点

（1）可构建在廉价机器上

① 通过多副本提高可靠性，提供了容错和恢复机制。

② 服务器节点的宕机场景是常态，采用多 NameNode 节点和多副本的机制可以有效防止单点故障。

（2）高容错性

① 数据自动保存多个副本，副本丢失后自动恢复。

② HDFS 的核心设计思想：大文件被切分为小文件，使用分而治之的思想让很多服务器对同一个文件进行联合管理；每个小文件做冗余备份，并且分散存储到不同的服务器，做到高可靠不丢失。

（3）适合批处理

① 移动计算而非数据，数据位置暴露给计算框架。

② 海量数据的计算任务最终一定要被切分成很多的小任务进行。

（4）适合大数据处理

可用于处理吉字节、太字节甚至拍字节级数据，百万规模以上的文件数量，一万以上节点规模。

(5) 流式文件访问

一次性写入，多次读取，保证数据一致性。

2. HDFS 的缺点

(1) 不适合以下操作

① 低延迟数据访问，如毫秒级响应。

② 小文件存取。占用 NameNode 节点大量内存，造成数据寻道时间超过读取时间。

③ 并发写入、文件随机修改。一个文件只能有一个写入者，仅支持 append 操作。

(2) HDFS 为什么不适用于存储小文件？

这和 HDFS 系统底层设计实现有关，HDFS 本身的设计目的就是解决海量大文件数据的存储问题，HDFS"天生喜欢"大数据的处理。大文件存储在 HDFS 中，会被切分成很多小数据块，任何一个文件不管有多小，都是一个独立的数据块，而这些数据块的信息保存在元数据中，在之前的内容里，我们介绍过在 HDFS 集群的 NameNode 节点中会存储元数据的信息，这里再强调一下，元数据的信息主要包括以下 3 部分：

① 抽象目录树；

② 文件和数据块的映射关系，一个数据块的元数据大小约是 150 B；

③ 数据块的多个副本存储地。

元数据存储在磁盘(①和②)和内存(①、②和③)中，而服务器的内存是有上限的。例如：有 100 个 1 MB 的文件存入 HDFS，那么数据块的个数就是 100 个，元数据的大小就是 100×150 B，消耗了 15 000 B 的内存，但是只存储了 100 MB 的数据；有 1 个 100 MB 的文件存入 HDFS，那么数据块的个数就是 1 个，元数据的大小就是 150 B，消耗了 150 B 的内存，存储了 100 MB 的数据。所以说 HDFS 不适用于存储小文件。

2.3.2 HDFS 的辅助功能

HDFS 作为一个分布式文件系统，主要提供两个最重要的功能：文件上传和下载。而为了保障这两个功能的完美和高效实现，HDFS 提供了很多辅助功能。

1. 心跳机制

① Hadoop 集群是 Master/Slave 结构，Master 节点中有 NameNode 和 ResourceManager 进程，Slave 节点中有 DataNode 和 NodeManager 进程。

② Master 启动时会启动一个 IPC(Inter-Process Communication，进程间通信)Server 服务，等待 Slave 的链接。

③ Slave 启动时会主动链接 Master 的 IPC Server 服务，并且每隔 3 秒链接一次 Master 节点，间隔时间是可以调整的，参数为 dfs.heartbeat.interval，这个每隔一段时间链接一次的机制，我们形象地称为心跳。Slave 通过心跳汇报自己的信息给 Master，Master 通过心跳给 Slave 下达命令。

④ NameNode 通过心跳得知 DataNode 的状态，ResourceManager 通过心跳得知 NodeManager 的状态。

⑤ 如果 Master 长时间没有收到 Slave 的心跳，NameNode 会向 DataNode 确认 2 次，每 5 分钟确认一次。如果 2 次都没有返回结果，就认为该 Slave 节点"挂掉"了。一般 NameNode 判断一个 DataNode 失效的时间会有一个计算公式：

第 2 章 Hadoop 分布式文件系统

timeout＝10×心跳间隔时间＋2×检查一次消耗的时间

正常情况下,心跳间隔时间 dfs.heartbeat.interval 为 3 s,检查一次消耗的时间 heartbeat.recheck.interval 为 5 min,最终判断节点失效时间为 630 s。节点之间可以查看最后联系的时间,如图 2-4 所示。

图 2-4　节点的最后联系时间

2. 安全模式

① HDFS 的启动(关闭)是先启动(关闭)NameNode 节点,再启动(关闭)DataNode 节点,最后启动(关闭)SecondaryNameNode 节点。

② 决定 HDFS 集群启动时长的因素一般是:
- 磁盘元数据的大小;
- DataNode 节点的个数。

当元数据很大或者节点个数很多的时候,HDFS 集群的启动需要一段很长的时间,那么在还没有完全启动的时候 HDFS 能否对外提供服务？在 HDFS 的启动命令 start-dfs.sh 执行的时候,HDFS 会自动进入安全模式。为了确保用户的操作是可以高效执行成功的,HDFS 在发现自身不完整的时候,会进入安全模式来保护自己。启动过程中,汇总信息中会有图 2-5 所示提示。

在正常启动之后,如果 HDFS 发现所有的数据都是齐全的,那么 HDFS 会退出安全模式,如图 2-6 所示。

③ 对安全模式进行测试。

安全模式常用操作命令:

```
hdfs dfsadmin -safemode leave  //强制 NameNode 退出安全模式
hdfs dfsadmin -safemode enter  //进入安全模式
hdfs dfsadmin -safemode get    //查看安全模式状态
hdfs dfsadmin -safemode wait   //等待,一直到安全模式结束
```

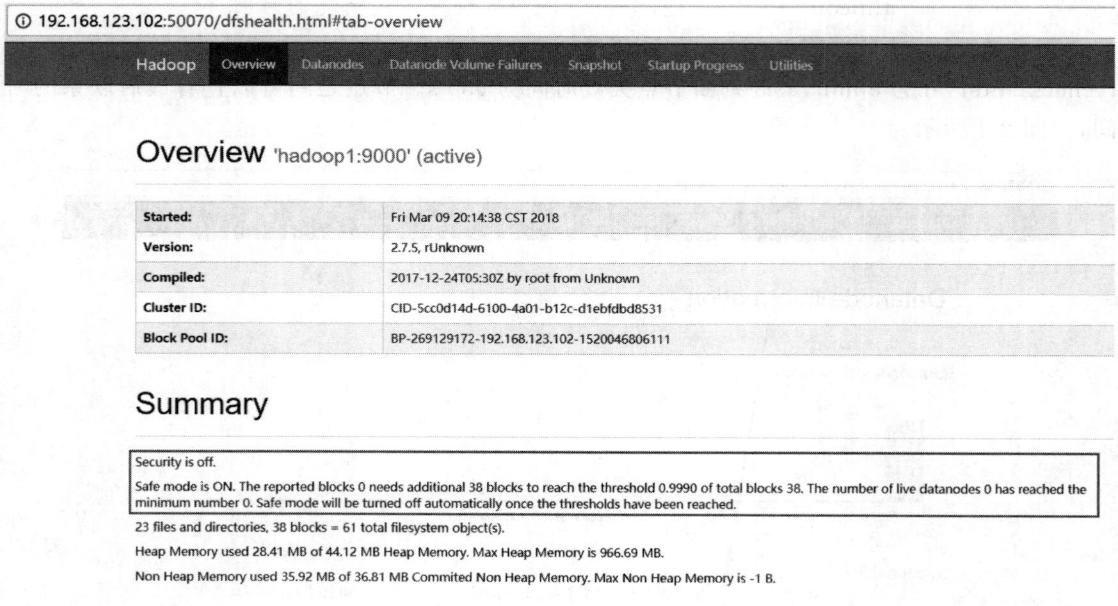

图 2-5 安全模式的提示

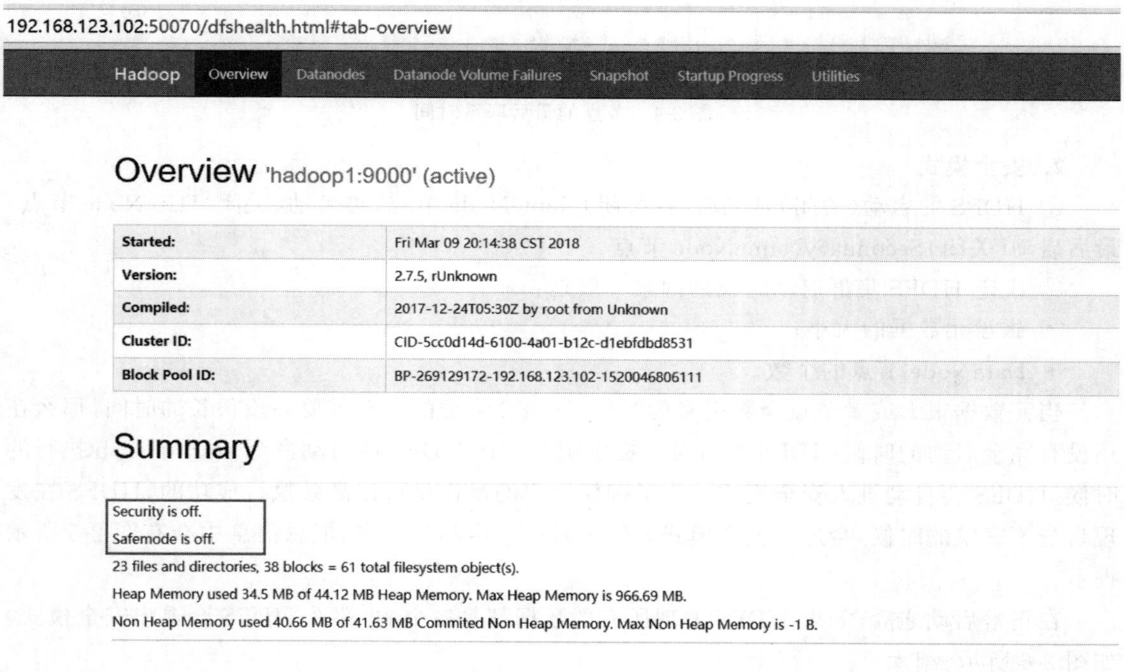

图 2-6 退出安全模式的提示

手工进入安全模式进行测试。测试创建文件夹：

```
[hadoop@hadoop1 ~]$ hdfs dfsadmin -safemode enter
Safe mode is ON
[hadoop@hadoop1 ~]$ hadoop fs -mkdir -p /xx/yy/zz
mkdir: Cannot create directory /xx/yy/zz. Name node is in safe mode.
```

测试下载文件：

```
[hadoop@hadoop1 ~]$ ls
apps    data
[hadoop@hadoop1 ~]$ hdfs dfsadmin -safemode get
Safe mode is ON
[hadoop@hadoop1 ~]$ hadoop fs -get /aa/1.txt ~/1.txt
[Hadoop@Hadoop1 ~]$ ls
1.txt   apps    data
```

测试上传文件：

```
[hadoop@hadoop1 ~]$ hadoop fs -put 1.txt /a/xx.txt
put: Cannot create file/a/xx.txt._COPYING_. Name node is in safe mode.
```

通过以上测试结果，我们可以得出结论，在安全模式下，HDFS 集群文件系统具备如下特征：

- 如果一个操作涉及元数据的修改，则不能进行操作。
- 如果一个操作仅仅是查询，则是被允许的。
- 所谓的安全模式，只是保护 NameNode 节点，而不是保护 DataNode 节点。

3. 副本存放策略

通常来说，大型的 Hadoop 集群是以机架的形式来组织的，它们分布在不同的机架上，同一个机架的节点往往通过同一个网络交换机连接，在网络带宽方面与跨机架通信相比有较大优势。但是如果某一个文件数据库同时存储在同一个机架上，可能由于各种故障，导致文件不可用。HDFS 采用机架感知策略来改进数据的可靠性、高可用性和网络带宽的利用率。

通过机架感知过程，NameNode 节点可以确定每一个 DataNode 节点所属的机架 ID。一个简单但没有优化的策略就是将副本存放在不同的机架上，这样可以防止整个机架失效时数据丢失，并且允许读取数据的时候充分利用多个机架的带宽。这种设置策略可以将副本均匀分布在集群中，有利于组件失效情况下的数据均衡。

HDFS 会尽量让读取任务去读取距离客户端最近的副本数据来减少整体带宽消耗，从而降低整体的带宽延时。

通常而言，一个机架共享一个电源、一条网线、一个交换机，HDFS 备份通常在同一个机架上存储一份，在另外一个机架上存储两份。

通过机架感知，处于工作状态的 HDFS 总是设法确保数据块的 3 个（或者更多）副本中至少有两个处在同一机架，至少有一个处在不同机架。

而 HDFS 为每一个数据块存 3 个副本的话，那么客户端如何写入呢？一般会采取如下存储策略，如图 2-7 所示。

- 第一个副本：放置在上传文件的 DataNode 上；如果是集群外提交，则随机挑选一台磁盘不太慢、CPU 不太忙的节点。
- 第二个副本：放置在与第一个副本不同的机架的节点上。
- 第三个副本：放置在与第二个副本相同的机架的不同节点上。
- 更多的副本：随机放在节点上。

4. 数据均衡

数据均衡理想状态：节点均衡、机架均衡和磁盘均衡。Hadoop 的 HDFS 集群非常容易出

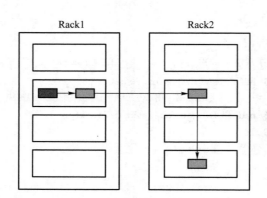

图 2-7 副本摆放策略

现机器与机器之间磁盘利用率不平衡的情况,例如:集群内新增、删除节点,或者某个节点机器内硬盘存储达到饱和值。当数据不平衡时,Map 任务可能会被分配到没有存储数据的机器上,这将导致网络带宽的消耗,也无法很好地进行本地计算。

当 HDFS 数据不均衡时,需要对 HDFS 进行数据的负载均衡调整,即对各节点机器上数据的存储分布进行调整,从而让数据均匀地分布在各个 DataNode 上,均衡 I/O 性能,防止热点的发生。进行数据的负载均衡调整时,必须满足如下原则:

- 数据均衡不能导致数据块减少、数据块备份丢失。
- 管理员可以中止数据均衡进程。
- 每次移动的数据量以及占用的网络资源必须是可控的。
- 数据均衡过程不能影响 NameNode 的正常工作。

(1) 数据均衡的原理

数据均衡过程的核心是一个数据均衡算法,该数据均衡算法将不断迭代数据均衡逻辑,直至集群内数据均衡为止。该数据均衡算法每次迭代的逻辑如图 2-8 所示。

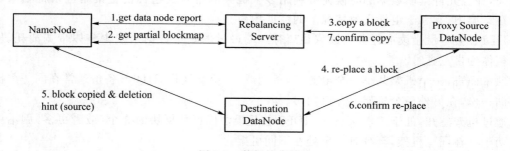

图 2-8 数据均衡算法

步骤分析如下:

① 数据均衡服务(Rebalancing Server)首先要求 NameNode 生成 DataNode 数据分布分析报告,获取每个 DataNode 的磁盘使用情况。

② Rebalancing Server 汇总需要移动的数据分布情况,计算具体数据块迁移路线图。数据块迁移路线图要确保是网络内最短路径。

③ 开始数据块迁移任务,Proxy Source DataNode 复制一块需要移动的数据块。

④ 将复制的数据块复制到目标 DataNode 上。

⑤ 删除原始数据块。

⑥ 目标 DataNode 向 Proxy Source DataNode 确认该数据块迁移完成。
⑦ Proxy Source DataNode 向 Rebalancing Server 确认本次数据块迁移完成。然后继续执行这个过程,直至集群达到数据均衡标准。

(2) DataNode 分组

在当前步骤中,HDFS 会把当前的 DataNode 节点根据阈值的设定情况划分到 Over、Above、Below、Under 4 个组中。在移动数据块的时候,Over 组、Above 组中的块向 Below 组、Under 组移动。4 个组的定义如图 2-9 所示。

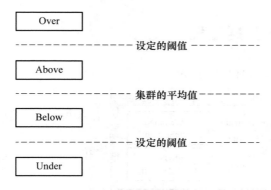

图 2-9　阈值的设定情况

(3) HDFS 数据自动平衡脚本的使用方法

在 Hadoop 集群中,包含一个 start-balancer.sh 脚本,通过运行这个工具可启动 HDFS 数据均衡服务。该工具可以做到热插拔,即无须重启计算机和 Hadoop 服务。节点 Hadoop_home/bin 目录下的 start-balancer.sh 脚本就是数据均衡服务的启动脚本,启动命令为"Hadoop_home/bin/start-balancer.sh -threshold"。通常影响 Balancer 的有以下几个参数。

① -threshold:
- 默认设置:10。参数取值范围:0~100。
- 参数含义:判断集群是否平衡的阈值。理论上,该参数设置得越小,整个集群就越平衡。

② dfs.balance.bandwidthPerSec:
- 默认设置:1048576(1 MB/s)。
- 参数含义:Balancer 运行时允许占用的带宽。

接下来,我们来演示一个案例,设置数据均衡,如下:

```
#启动数据均衡,默认阈值为 10%
$ Hadoop_home/bin/start-balancer.sh
#启动数据均衡,阈值为 5%
bin/start-balancer.sh -threshold 5
#停止数据均衡
$ Hadoop_home/bin/stop-balancer.sh
```

在 hdfs-site.xml 文件中可以设置数据均衡占用的网络带宽限制,设置如下:

```
<property>
    <name>dfs.balance.bandwidthPerSec</name>
    <value>1048576</value>
```

< description > Specifies the maximum bandwidth that each datanode can utilize for the balancing purpose in term of the number of bytes per second. </description>
</property>

2.4 HDFS 读写过程

2.4.1 HDFS 写入数据过程

HDFS 写入数据过程如图 2-10 所示。

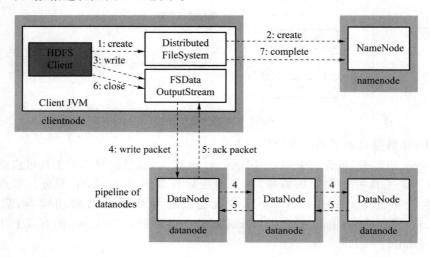

图 2-10 HDFS 写入数据过程

HDFS 是 Hadoop 的重要组件之一，对其进行数据的读写是很常见的操作，然而读者真的了解其读写过程吗？接下来我们一起来了解 HDFS 的数据写入过程部署，简述为如下步骤。

① 使用 HDFS 提供的客户端 Client 向远程的 NameNode 节点发起远程过程调用（RPC）请求。

② NameNode 节点会检查要创建的文件是否已经存在，创建者是否有权限进行操作，成功则会为文件创建一个记录，否则会让客户端抛出异常。

③ 当客户端开始写入文件的时候，客户端会将文件切分成多个 packets，并在内部以 data queue（数据队列）的形式管理这些 packets，并向 NameNode 节点申请数据块，获取用来存储副本的合适的 DataNode 节点列表，列表的大小根据 NameNode 节点中 replication 参数的设定而定。

④ 开始以 pipeline（管道）的形式将 packet 写入所有的副本中。客户端把 packet 以流的方式写入第一个 DataNode 节点，该 DataNode 节点存储该 packet 之后，再将其传递给此 pipeline 中的下一个 DataNode 节点，直到传递给最后一个 DataNode 节点，这种写数据的方式呈流水线的形式。

⑤ 最后一个 DataNode 节点成功存储之后会返回一个 ack packet（确认队列），其在

pipeline 中传递至客户端,在客户端的开发库内部维护着 data queue,成功收到 DataNode 返回的 ack packet 后会从 data queue 中移除相应的 packet。

⑥ 如果传输过程中某个 DataNode 节点出现了故障,那么当前的 pipeline 会被关闭,出现故障的 DataNode 节点会从当前的 pipeline 中移除,剩余的数据块会继续在剩下的 DataNode 节点中以 pipeline 的形式传输,同时,NameNode 节点会分配一个新的 DataNode 节点,保持设定的副本数量。

⑦ 客户端完成数据的写入后,会对数据流调用 close() 方法,关闭数据流。

⑧ 只要写入了 dfs.replication.min(最小写入成功的副本数,默认为 1)个副本,写操作就会成功,并且这个块可以在集群中异步复制,直到达到其目标副本数(dfs.replication 的默认值为 3)。因为 NameNode 已经知道文件由哪些块组成,所以它在返回成功前只需要等待数据块进行最小量的复制。

2.4.2 HDFS 读取数据过程

HDFS 读取数据过程如图 2-11 所示。

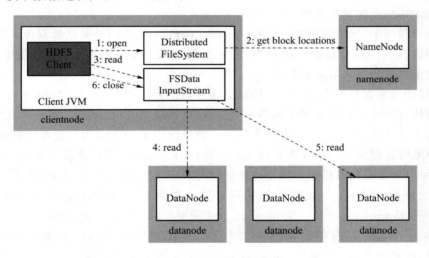

图 2-11　HDFS 读取数据过程

通过学习 Hadoop 官网的资料,我们整理出读取文件的步骤如下。

① 客户端调用 FileSystem 实例的 open() 方法,获得这个文件对应的输入流 DFSInputStream。

② 通过 RPC 远程调用 NameNode 节点,获得 NameNode 节点中这个文件对应的数据块的保存位置,包括这个文件的副本的保存位置(主要是各 DataNode 节点的地址)。

③ 获得输入流之后,客户端调用 read() 方法读取数据。选择最近的 DataNode 节点建立连接并读取数据。NameNode 节点返回保存该数据块的 DataNode 节点地址,同时根据距离客户端的远近对 DataNode 节点进行排序,然后 DistributedFileSystem 会利用 DFSInputStream 来实例化 FSDataInputStream 并返回给客户端,同时返回数据块对应的 DataNode 节点的地址。

④ 如果客户端和其中一个 DataNode 节点位于同一机器上(如 MapReduce 过程中的 Mapper 和 Reducer),那么会直接从本地读取数据。

⑤ 到达数据块末端,关闭与这个 DataNode 节点的连接,然后重新查找下一个数据块。
⑥ 不断执行第②～⑤步,直到数据全部读完。
⑦ 客户端调用 close()方法,关闭输入流 DFSInputStream。

2.5 分布式集群中 HDFS 的各种角色

2.5.1 NameNode 的可靠性

1. 问题场景

① NameNode 节点的磁盘故障导致 NameNode 服务宕机,如何挽救集群及数据?
② NameNode 节点是否可以有多个? NameNode 节点内存要配置为多大? NameNode 节点和集群数据存储能力有关系吗?
③ HDFS 分布式集群中文件的 blocksize 参数究竟设置为多大,才能达到最佳性能?

2. NameNode 元数据的管理

在计算机科学中,预写式日志(Write-Ahead Logging,WAL)是关系数据库系统中用于提供原子性和持久性(ACID 属性中的两个)的一系列技术。在使用 WAL 的系统中,所有的修改在提交之前都要先写入 log 文件中。

log 文件中通常包括 redo 和 undo 信息。这样做的目的可以通过一个例子来说明。假设在一个程序执行某些操作的过程中机器掉电了,在重新启动时,程序可能需要知道当时执行的操作是成功了还是部分成功了,或者是失败了。如果使用了 WAL,程序就可以检查 log 文件,并对突然掉电时计划执行的操作内容和实际上执行的操作内容进行比较。在这个比较的基础上,程序就可以决定是撤销已做的操作还是继续完成已做的操作,或者是保持原样。

WAL 允许用 in-place 方式更新数据库。另一种用来实现原子更新的方法是 shadow paging,它并不是 in-place 方式。用 in-place 方式做更新的主要优点是减少索引和块列表的修改。ARIES 是 WAL 系列技术常用的算法,在文件系统中 WAL 通常称为 journaling。PostgreSQL 也是用 WAL 来提供 point-in-time 恢复和数据库复制特性。

NameNode 节点对数据的管理采用了两种存储形式:内存和磁盘。

首先,内存中存储了一份完整的元数据(metadata),包括目录树结构、文件、数据块和副本存储地的映射关系。其次,磁盘中也存储了一份完整的元数据。磁盘元数据镜像文件,如 fsimage_0000000000000000555,内容等价于 edits_0000000000000000001-0000000000000000018 ……edits_0000000000000000444-0000000000000000555 文件内容合并。

数据历史操作日志文件 edits:edits_0000000000000000001-0000000000000000018(可通过日志运算出元数据,全部存储在磁盘中)。

数据预写操作日志文件 edits_inprogress_0000000000000000556(存储在磁盘中)。

3. NameNode 元数据存储机制

① 内存中有一份完整的元数据。
② 磁盘中有一个"准完整"的元数据镜像文件 fsimage(在 NameNode 的工作目录中)。
③ 用于衔接内存 metadata 和持久化元数据镜像 fsimage 的操作日志(edits 文件),当客

户端对 HDFS 中的文件执行新增或者修改操作时,操作记录首先被记入 edits 文件中,当客户端操作成功后,相应的元数据会更新到内存 metadata 中。

2.5.2 DataNode 的可靠性

1. 问题场景

① Hadoop 集群容量不够时怎么扩容？
② 如果有一些 DataNode 节点宕机,该怎么办？
③ DataNode 节点已启动,但是集群的可用 DataNode 列表中没有显示,该怎么办？

2. DataNode 死亡判断时限参数

DataNode 进程死亡或者网络故障造成 DataNode 无法与 NameNode 通信时,NameNode 不会立即把该节点判定为死亡,要经过一段时间,这段时间暂称作超时时长。HDFS 默认的超时时长为 10 min+30 s。如果定义超时时长为 timeout,则超时时长的计算公式为

$$timeout = 2 \times heartbeat.recheck.interval + 10 \times dfs.heartbeat.interval$$

heartbeat.recheck.interval 默认为 5 min,dfs.heartbeat.interval 默认为 3 s。需要注意的是,hdfs-site.xml 配置文件中的 heartbeat.recheck.interval 的单位为毫秒,dfs.heartbeat.interval 的单位为秒。举个例子,如果 heartbeat.recheck.interval 设置为 5 000 ms,dfs.heartbeat.interval 设置为 3 s(默认),则总的超时时长为 40 s。

这些时限参数可以通过配置文件进行修改,参考配置如下:

```
<property>
<!--HDFS 集群数据冗余块的自动删除时长,单位为毫秒,默认为 1 h-->
<name>dfs.blockreport.intervalMsec</name>
<value>3600000</value>
<description>Determines block reporting interval in milliseconds.</description>
</property>
<property>
<name>heartbeat.recheck.interval</name>
<value>5000</value>
</property>
<property>
<name>dfs.heartbeat.interval</name>
<value>3</value>
</property>
```

2.5.3 元数据的 CheckPoint

每隔一段时间,SecondaryNameNode 节点会将 NameNode 节点上积累的所有 edits 和一个最新的 fsimage 下载到本地,并加载到内存进行合并,这个过程称为 CheckPoint。

1. CheckPoint 触发配置

我们通过修改配置文件进行 CheckPoint 参数配置,参考配置和说明如下:

```
dfs.namenode.checkpoint.check.period = 60
＃检查触发条件是否满足的频率,60 s
dfs.namenode.checkpoint.dir = file://${hadoop.tmp.dir}/dfs/namesecondary
＃以上两个参数做 CheckPoint 操作时,SecondaryNameNode 的本地工作目录
dfs.namenode.checkpoint.edits.dir = ${dfs.namenode.checkpoint.dir}
dfs.namenode.checkpoint.max-retries = 3  ＃＃最大重试次数
dfs.namenode.checkpoint.period = 3600  ＃＃两次 CheckPoint 之间的时间间隔,3600 s
dfs.namenode.checkpoint.txns = 1000000  ＃＃两次 CheckPoint 之间最大的操作记录
```

2. CheckPoint 的其他功能

NameNode 节点和 SecondaryNameNode 节点的工作目录存储结构完全相同,所以当 NameNode 节点故障退出需要重新恢复时,可以从 SecondaryNameNode 节点的工作目录中将 fsimage 复制到 NameNode 节点的工作目录,以恢复 NameNode 节点的元数据。

本 章 小 结

1. NameNode 节点的工作职责

NameNode 管理文件系统的命名空间,它维护着文件系统树及整棵树上所有的文件和目录。这些信息以两个文件的形式永久保存在本地磁盘上:命名空间镜像文件和编辑日志文件。NameNode 也记录着每个文件的每个数据块所在的数据节点信息,但它并不永久保存块的位置信息,因为这些信息在系统启动时会由数据节点重建。

2. DataNode 节点的工作职责

DataNode 节点用来存储管理用户的文件数据块,同时通过心跳信息定期向 NameNode 汇报自身所持有的数据块信息。

3. SecondaryNameNode 节点的工作职责

SecondaryNameNode 节点的作用就是分担 NameNode 节点合并元数据的压力。生产环境中,在配置 SecondaryNameNode 节点的工作节点时,建议不要和 NameNode 节点处于同一节点。事实上,只有在普通的伪分布式集群和分布式集群中才会有 SecondaryNameNode 节点这个角色,在 Hadoop HA 或者联邦集群中都不再出现该角色。在 Hadoop HA 和联邦集群中,都是由 StandbyNameNode 节点承担 NameNode 节点的灾备能力。

第 3 章

MapReduce 并行计算框架

3.1　MapReduce 概述

MapReduce 是一个分布式运算程序的编程框架，是用户开发"基于 Hadoop 的数据分析应用"的核心框架。MapReduce 的核心功能是将用户编写的业务逻辑代码和自带默认组件整合成一个完整的分布式运算程序，并发运行在一个 Hadoop 分布式集群上。

3.1.1　为什么需要 MapReduce？

在单机上处理海量数据时，因为硬件资源限制，当程序由单机版本运行环境扩展为分布式环境时，将极大增加程序的复杂度和开发难度，会引入大量的复杂工作。为了提高开发效率，可以将分布式程序中的公共功能封装成框架，让程序开发人员可以将精力集中于业务逻辑。

Hadoop 生态中 MapReduce 就是这样的一个分布式程序运算框架，它把大量分布式程序都会涉及的内容封装进来，让用户可以专注于自己的业务逻辑代码的开发。引入 MapReduce 框架后，开发人员可以将绝大部分工作集中在业务逻辑的开发上，而将分布式计算中的复杂性问题交由框架来处理。MapReduce 进行并发计算时，对应以上问题的整体处理结构可以分为以下 3 个核心部分。

- MRAppMaster：MapReduce Application Master，分配任务、协调任务的运行。
- MapTask(YARNChild)：阶段并发任务，负责 Map 阶段的任务处理。
- ReduceTask(YARNChild)：阶段汇总任务，负责 Reduce 阶段的任务处理。

简单地讲，MapReduce 可以实现大数据并行计算处理。所谓大数据处理，即以价值为导向，对大数据进行加工、挖掘和优化等各种处理。MapReduce 擅长处理大数据，它为什么具有这种能力呢？这可以由 MapReduce 的设计思想来解释，MapReduce 的设计思想就是"分而治之"，具体工作可以理解成图 3-1 所示的处理流程。

① Mapper 负责拆分，即把复杂的任务分解为若干个简单的任务来处理。"简单的任务"包含 3 层含义：一是数据或计算的规模相对于原始任务要大大缩小；二是就近计算原则，即任务会被分配到存放着所需数据的节点上进行计算；三是这些小任务可以并行计算，彼此间几乎没有依赖关系。

② Reducer 负责对 Map 阶段的结果进行汇总。至于需要多少个 Reducer，用户可以根据具体问题，在 mapred-site.xml 配置文件中设置参数 mapred.reduce.tasks 的值，默认值为 1。

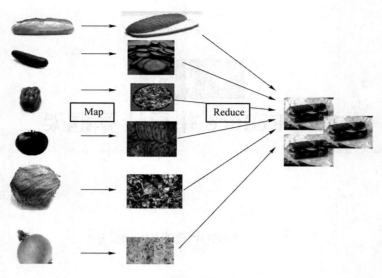

图 3-1　并行计算处理流程

3.1.2　MapReduce 程序运行演示

在 Hadoop MapReduce 组件里，官方提供了一些示例程序，其中非常有名的就是 WordCount 和 Pi 程序。这些 MapReduce 程序的代码都在 hadoop-mapreduce-examples-2.7.5.jar 包里，这个 jar 包在 Hadoop 安装目录下的 /share/hadoop/mapreduce/ 目录里，下面我们使用 hadoop 命令来试跑示例程序，看看运行效果。

1. MapReduce 示例：Pi 程序

```
[hadoop@hadoop1 ~]$ cd apps/hadoop-2.7.5/share/hadoop/mapreduce/
[hadoop@hadoop1 mapreduce]$ pwd
/home/hadoop/apps/hadoop-2.7.5/share/hadoop/mapreduce
[hadoop@hadoop1 mapreduce]$ hadoop jar hadoop-mapreduce-examples-2.7.5.jar pi 5 5
```

部分输出结果参考如下：

```
File System Counters
        FILE: Number of bytes read = 1655738
        FILE: Number of bytes written = 3342014
        FILE: Number of read operations = 0
        FILE: Number of large read operations = 0
        FILE: Number of write operations = 0
        HDFS: Number of bytes read = 2360
        HDFS: Number of bytes written = 3755
        HDFS: Number of read operations = 87
        HDFS: Number of large read operations = 0
        HDFS: Number of write operations = 45
```

```
        Map-Reduce Framework
                Map input records = 5
                Map output records = 10
                Map output bytes = 90
                Map output materialized bytes = 140
                Input split bytes = 735
                Combine input records = 0
                Combine output records = 0
                Reduce input groups = 2
                Reduce shuffle bytes = 140
                Reduce input records = 10
                Reduce output records = 0
                Spilled Records = 20
                Shuffled Maps = 5
                Failed Shuffles = 0
                Merged Map outputs = 5
                GC time elapsed (ms) = 684
                Total committed heap usage (bytes) = 1257725952
        Shuffle Errors
                BAD_ID = 0
                CONNECTION = 0
                IO_ERROR = 0
                WRONG_LENGTH = 0
                WRONG_MAP = 0
                WRONG_REDUCE = 0
        File Input Format Counters
                Bytes Read = 590
        File Output Format Counters
                Bytes Written = 97
Job Finished in 17.524 seconds
Estimated value of Pi is 3.68000000000000000000
```

2. MapReduce 示例：WordCount 程序

[hadoop@hadoop1 mapreduce]$ hadoop jar hadoop-mapreduce-examples-2.7.5.jar wordcount /wc/input1/ /wc/output1/

示例中使用 WordCount 来统计分布式文件夹/wc/input1/，统计结果放到/wc/output1/目录下，输出结果如下：

```
File System Counters
    FILE: Number of bytes read = 889648
    FILE: Number of bytes written = 1748828
    FILE: Number of read operations = 0
    FILE: Number of large read operations = 0
    FILE: Number of write operations = 0
Map-Reduce Framework
    Map input records = 3
```

```
        Map output records = 18
        Map output bytes = 179
        Map output materialized bytes = 174
        Input split bytes = 256
        Combine input records = 18
        Combine output records = 14
        Reduce input groups = 11
        Reduce shuffle bytes = 174
        Reduce input records = 14
        Reduce output records = 11
        Spilled Records = 28
        Shuffled Maps = 2
        Failed Shuffles = 0
        Merged Map outputs = 2
        GC time elapsed (ms) = 43
        Total committed heap usage (bytes) = 457912320
    Shuffle Errors
        BAD_ID = 0
        CONNECTION = 0
        IO_ERROR = 0
        WRONG_LENGTH = 0
        WRONG_MAP = 0
        WRONG_REDUCE = 0
    File Input Format Counters
        Bytes Read = 107
    File Output Format Counters
        Bytes Written = 94
```

查看统计结果如下,具体结果取决于读者的文件内容,本书中的结果仅供参考:

```
[hadoop@hadoop1 mapreduce]$ hadoop fs -cat /wc/output1/part-r-00000
80    1
82    1
90    2
92    2
94    1
95    1
96    1
98    1
99    1
```

3.1.3 WordCount.java 源码分析

WordCount.java 源码如下:

```java
/**
 * Licensed to the Apache Software Foundation (ASF) under one
 * or more contributor license agreements.  See the NOTICE file
 * distributed with this work for additional information
 * regarding copyright ownership.  The ASF licenses this file
 * to you under the Apache License, Version 2.0 (the
 * "License"); you may not use this file except in compliance
 * with the License.  You may obtain a copy of the License at
 *
 *     http://www.apache.org/licenses/LICENSE-2.0
 *
 * Unless required by applicable law or agreed to in writing, software
 * distributed under the License is distributed on an "AS IS" BASIS,
 * WITHOUT WARRANTIES OR CONDITIONS OF ANY KIND, either express or implied.
 * See the License for the specific language governing permissions and
 * limitations under the License.
 */
package org.apache.hadoop.examples;
import java.io.IOException;
import java.util.StringTokenizer;

import org.apache.hadoop.conf.Configuration;
import org.apache.hadoop.fs.Path;
import org.apache.hadoop.io.IntWritable;
import org.apache.hadoop.io.Text;
import org.apache.hadoop.MapReduce.Job;
import org.apache.hadoop.MapReduce.Mapper;
import org.apache.hadoop.MapReduce.Reducer;
import org.apache.hadoop.MapReduce.lib.input.FileInputFormat;
import org.apache.hadoop.MapReduce.lib.output.FileOutputFormat;
import org.apache.hadoop.util.GenericOptionsParser;
public class WordCount {
  public static class TokenizerMapper
       extends Mapper<Object, Text, Text, IntWritable>{
    private final static IntWritable one = new IntWritable(1);
    private Text word = new Text();
    public void map(Object key, Text value, Context context
                    ) throws IOException, InterruptedException {
      StringTokenizer itr = new StringTokenizer(value.toString());
      while (itr.hasMoreTokens()) {
        word.set(itr.nextToken());
        context.write(word, one);
      }
    }
  }
  public static class IntSumReducer
```

```java
            extends Reducer<Text,IntWritable,Text,IntWritable> {
    private IntWritable result = new IntWritable();
    public void reduce(Text key, Iterable<IntWritable> values, Context context
                      ) throws IOException, InterruptedException {
      int sum = 0;
      for (IntWritable val : values) {
        sum += val.get();
      }
      result.set(sum);
      context.write(key, result);
    }
  }
  public static void main(String[] args) throws Exception {
    Configuration conf = new Configuration();
    String[] otherArgs = new GenericOptionsParser(conf, args).getRemainingArgs();
    if (otherArgs.length < 2) {
      System.err.println("Usage: wordcount <in> [<in>...] <out>");
      System.exit(2);
    }
    Job job = Job.getInstance(conf, "word count");
    job.setJarByClass(WordCount.class);
    job.setMapperClass(TokenizerMapper.class);
    job.setCombinerClass(IntSumReducer.class);
    job.setReducerClass(IntSumReducer.class);
    job.setOutputKeyClass(Text.class);
    job.setOutputValueClass(IntWritable.class);
    for (int i = 0; i < otherArgs.length - 1; ++i) {
      FileInputFormat.addInputPath(job, new Path(otherArgs[i]));
    }
    FileOutputFormat.setOutputPath(job, new Path(otherArgs[otherArgs.length - 1]));
    System.exit(job.waitForCompletion(true) ? 0 : 1);
  }
}
```

1. 源码分析

由 WordCount 这个 MapReduce 程序的源码可以得出以下几点结论:

- 该程序有一个 main()方法,用来启动任务的运行,其中 Job 对象存储了该程序运行的必要信息,如指定 Mapper 类〔job.setMapperClass(TokenizerMapper.class)〕和 Reducer 类〔job.setReducerClass(IntSumReducer.class)〕。
- 该程序中的 TokenizerMapper 类继承了 Mapper 类。
- 该程序中的 IntSumReducer 类继承了 Reducer 类。

2. WordCount 工作原理

理解了 MapReduce 的工作原理,那么就从一个 Job 开始,从分 Map 任务和 Reduce 任务开始。用户编写的程序分为 3 个部分:Mapper、Reducer、Driver。

Mapper 的输入数据和输出数据是 KV 对(键值对)的形式(KV 的类型可自定义),

Mapper 的业务逻辑写在 map()方法中,map()方法(MapTask 进程)对每一组<k,v>调用一次。

Reducer 的输入数据类型对应 Mapper 的输出数据类型,也是 KV 对。Reducer 的业务逻辑写在 reduce()方法中,reduce()方法对相同的<k,v>组调用一次。

用户的 Mapper 和 Reducer 都要继承各自的父类。整个程序需要一个 Driver 来进行提交,提交的是一个描述了各种必要信息的 Job 对象。

3.1.4 编写自己的 WordCount 程序

1. 准备数据文件,并上传到 HDFS 上,路径:/input/wordcount.txt

wordcount.txt 内容如下:

```
Hello Hadoop
Hello BigData
Hello Spark
Hello Flume
Hello Kafka
```

2. 编写 WordCount 代码

这里用户可以输入 3 个参数,分别为应用的名称、数据文件的路径、结果的输出路径。

```java
[hadoop@hadoop1 wordcount]$ vim wordcountpara.java
package ls.wordcount;
import org.apache.hadoop.conf.Configuration;
import org.apache.hadoop.fs.Path;
import org.apache.hadoop.io.IntWritable;
import org.apache.hadoop.io.Text;
import org.apache.hadoop.mapreduce.Job;
import org.apache.hadoop.mapreduce.Mapper;
import org.apache.hadoop.mapreduce.Reducer;
import org.apache.hadoop.mapreduce.lib.input.FileInputFormat;
import org.apache.hadoop.mapreduce.lib.input.NLineInputFormat;
import org.apache.hadoop.mapreduce.lib.output.FileOutputFormat;

import java.io.IOException;
import java.util.StringTokenizer;

public class WordCount {

    public static class TokenizerMapper extends Mapper<Object, Text, Text, IntWritable> {

        private final static IntWritable one = new IntWritable(1);
        private Text word = new Text();
```

```java
        public void map(Object key, Text value, Context context)
            throws IOException, InterruptedException {
          StringTokenizer itr = new StringTokenizer(value.toString());
          while (itr.hasMoreTokens()) {
            word.set(itr.nextToken());
            context.write(word, one);
          }
        }
    }
    public static class IntSumReducer extends Reducer<Text, IntWritable, Text, IntWritable> {
        private IntWritable result = new IntWritable();

        public void reduce(Text key, Iterable<IntWritable> values, Context context)
            throws IOException, InterruptedException {
          int sum = 0;
          for (IntWritable val : values) {
            sum += val.get();
          }
          result.set(sum);
          context.write(key, result);
        }
    }

    public static void main(String[] args) throws Exception {
      if (args == null || args.length < 3) {
          args[0] = "wordcount";
          args[1] = "/input/wordcount.txt";
          args[2] = "/output/wordcountpara1";
      }
        Configuration conf = new Configuration();
        Job job = Job.getInstance(conf, args[0]);
        job.setJarByClass(WordCount.class);
        job.setMapperClass(TokenizerMapper.class);
        job.setCombinerClass(IntSumReducer.class);
        job.setReducerClass(IntSumReducer.class);
        job.setOutputKeyClass(Text.class);
        job.setOutputValueClass(IntWritable.class);
        job.setInputFormatClass(NLineInputFormat.class);
        // 输入文件路径
        FileInputFormat.addInputPath(job, new Path(args[1]));
        // 输出文件路径
        FileOutputFormat.setOutputPath(job, new Path(args[2]));
        System.exit(job.waitForCompletion(true) ? 0 : 1);
    }
}
```

3. 打包成 jar 包并上传至服务器本地(不需要上传到 HDFS 上)

在 maven 上进行 package,在 pom.xml 中加入下面的内容,主要是添加主类的入口:

```xml
<mainClass>ls.wordcount.WordCount</mainClass>
<build>
    <plugins>
        <plugin>
            <groupId>org.apache.maven.plugins</groupId>
            <artifactId>maven-jar-plugin</artifactId>
            <configuration>
                <archive>
                    <manifest>
                        <mainClass>ls.wordcount.WordCount</mainClass>
                        <addClasspath>true</addClasspath>
                        <classpathPrefix>lib/</classpathPrefix>
                    </manifest>
                </archive>
                <classesDirectory>
                </classesDirectory>
            </configuration>
        </plugin>
    </plugins>
```

4. 运行 hadoop jar 包

```
[hadoop@hadoop1 wordcount]$ hadoop jar /home/hadoop/wordcount/ls-hadoop-1.0-SNAPSHOT.jar wordcountpara /input/wordcount.txt /output/wordcountpara1.txt
20/03/09 10:54:49 WARN util.NativeCodeLoader: Unable to load native-hadoop library for your platform... using builtin-java classes where applicable
20/03/09 10:54:50 INFO client.RMProxy: Connecting to ResourceManager at hadoop1/192.168.123.101:8032
20/03/09 10:54:51 WARN mapreduce.JobResourceUploader: Hadoop command-line option parsing not performed. Implement the Tool interface and execute your application with ToolRunner to remedy this.
20/03/09 10:54:52 INFO input.FileInputFormat: Total input paths to process : 1
20/03/09 10:54:52 INFO mapreduce.JobSubmitter: number of splits:5
20/03/09 10:54:52 INFO mapreduce.JobSubmitter: Submitting tokens for job: job_1536504640893_0001
20/03/09 10:54:53 INFO impl.YarnClientImpl: Submitted application application_1536504640893_0001
20/03/09 10:54:53 INFO mapreduce.Job: The url to track the job: http://node1:8088/proxy/application_1536504640893_0001/
20/03/09 10:54:53 INFO mapreduce.Job: Running job: job_1536504640893_0001
20/03/09 10:55:01 INFO mapreduce.Job: Job job_1536504640893_0001 running in uber mode : false
20/03/09 10:55:01 INFO mapreduce.Job:  map 0% reduce 0%
20/03/09 10:55:13 INFO mapreduce.Job:  map 20% reduce 0%
20/03/09 10:55:14 INFO mapreduce.Job:  map 40% reduce 0%
20/03/09 10:55:21 INFO mapreduce.Job:  map 60% reduce 0%
```

```
20/03/09 10:55:23 INFO mapreduce.Job:  map 100% reduce 0%
20/03/09 10:55:24 INFO mapreduce.Job:  map 100% reduce 100%
20/03/09 10:55:25 INFO mapreduce.Job: Job job_1536504640893_0001 completed successfully
20/03/09 10:55:25 INFO mapreduce.Job: Counters: 50
        File System Counters
                FILE: Number of bytes read = 129
                FILE: Number of bytes written = 711389
                FILE: Number of read operations = 0
                FILE: Number of large read operations = 0
                FILE: Number of write operations = 0
                HDFS: Number of bytes read = 669
                HDFS: Number of bytes written = 51
                HDFS: Number of read operations = 18
                HDFS: Number of large read operations = 0
                HDFS: Number of write operations = 2
        Job Counters
                Killed map tasks = 1
                Launched map tasks = 5
                Launched reduce tasks = 1
                Other local map tasks = 5
                Total time spent by all maps in occupied slots (ms) = 74163
                Total time spent by all reduces in occupied slots (ms) = 7315
                Total time spent by all map tasks (ms) = 74163
                Total time spent by all reduce tasks (ms) = 7315
                Total vcore-milliseconds taken by all map tasks = 74163
                Total vcore-milliseconds taken by all reduce tasks = 7315
                Total megabyte-milliseconds taken by all map tasks = 75942912
                Total megabyte-milliseconds taken by all reduce tasks = 7490560
        Map-Reduce Framework
                Map input records = 5
                Map output records = 10
                Map output bytes = 103
                Map output materialized bytes = 153
                Input split bytes = 485
                Combine input records = 10
                Combine output records = 10
                Reduce input groups = 6
                Reduce shuffle bytes = 153
                Reduce input records = 10
                Reduce output records = 6
                Spilled Records = 20
                Shuffled Maps = 5
                Failed Shuffles = 0
                Merged Map outputs = 5
                GC time elapsed (ms) = 1767
                CPU time spent (ms) = 3720
```

```
            Physical memory (bytes) snapshot = 886165504
            Virtual memory (bytes) snapshot = 2868084736
            Total committed heap usage (bytes) = 618680320
    Shuffle Errors
            BAD_ID = 0
            CONNECTION = 0
            IO_ERROR = 0
            WRONG_LENGTH = 0
            WRONG_MAP = 0
            WRONG_REDUCE = 0
    File Input Format Counters
            Bytes Read = 184
    File Output Format Counters
            Bytes Written = 51
```

5．查看输出结果

```
[hadoop@hadoop1 wordcount]$ hadoop fs -cat /output/wordcountpara1.txt/part-r-00000
BigData   1
Flume     1
Hadoop    1
Hello     5
Kafka     1
Spark     1
```

3.2 MapReduce 的核心运行机制

一个完整的 MapReduce 程序在分布式运行时有以下两类实例进程。
- MRAppMaster：负责整个程序的过程调度及状态协调。
- MapTask(YARNChild)：负责 Map 阶段的整个数据处理流程。
- ReduceTask(YARNChild)：负责 Reduce 阶段的整个数据处理流程。

MapTask 和 ReduceTask 进程都是 YARNChild，这并不是说 MapTask 和 ReduceTask 跑在同一个 YARNChild 进程里。

1．MapReduce 程序的运行

① 一个 MapReduce 程序启动的时候，最先启动的是 MRAppMaster，MRAppMaster 启动后根据本次 Job 的描述信息，计算出需要的 MapTask 实例数量，然后向集群申请启动相应数量的 MapTask 进程。

② MapTask 进程启动之后，根据给定的数据切片范围（哪个文件的哪个偏移量范围）进行数据处理，主体流程为：
- 利用客户指定的 InputFormat 来获取 RecordReader，读取数据，形成输入 KV 对。

- 将输入 KV 对传递给客户定义的 map() 方法，做逻辑运算，并将 map() 方法输出的 KV 对收集到缓存。
- 将缓存中的 KV 对按照 key 分区排序后不断溢写到磁盘文件中。

③ MRAppMaster 监控到所有 MapTask 进程任务完成之后，会根据客户指定的参数启动相应数量的 ReduceTask 进程，并告知 ReduceTask 进程要处理的数据范围（数据分区）。

④ ReduceTask 进程启动之后，根据 MRAppMaster 告知的待处理数据所在位置，从若干台 MapTask 进程所在机器上获取若干个 MapTask 输出结果文件，并在本地进行重新归并排序，然后令 key 相同的 KV 对为一组，调用客户定义的 reduce() 方法进行逻辑运算，并收集运算输出的 KV 对，然后调用客户指定的 OutputFormat 将结果数据输出到外部存储。

2. MapTask 的并行度

在 MapReduce 程序的运行中，并不是 MapTask 越多就越好，需要考虑数据量的多少以及机器的配置。如果数据量很少，可能任务的启动时间都远远超过数据的处理时间。同样，也不是越少越好。那么应该如何切分呢？

假设有一个 300 MB 的文件，它会在 HDFS 中被切成 3 块：0～128 MB、128～256 MB、256～300 MB，并被放置到不同的节点上。在 MapReduce 任务中，这 3 个 Block 会被分给 3 个 MapTask。MapTask 在任务切片时实际上也是分配一个范围，只是这个范围是逻辑上的概念，与 Block 的物理划分没有什么关系。但在实践过程中如果 MapTask 读取的数据不在运行的本机上，则必须通过网络进行数据传输，对性能的影响非常大。所以常常采取的策略是按照 Block 的存储切分 MapTask，使得每个 MapTask 尽可能读取本机的数据。如果一些 Block 非常小，也可以把多个小 Block 交给一个 MapTask。

所以 MapTask 的切分要看情况进行处理，默认的实现是按照 Block 大小进行切分。MapTask 的切分工作由客户端（即开发的 main 方法）负责，一个切片就对应一个 MapTask 实例。

3. MapTask 并行度的决定机制

一个 Job 的 Map 阶段并行度由客户端在提交 Job 时决定。而客户端对 Map 阶段并行度的规划的基本逻辑为：对待处理数据执行逻辑切片（即按照一个特定切片大小，将待处理数据划分成逻辑上的多个 Split），然后每一个 Split 分配一个 MapTask 并行实例处理。这段逻辑及形成的切片规划描述文件由 FileInputFormat 实现类的 getSplits() 方法完成。

① 如果 Job 的每个 MapTask 或者 ReduceTask 的运行时间都只有 30～40 s，就要减少该 Job 的 Map 或者 Reduce 数，每一个 Task（Map 或者 Reduce）的启动和加入调度器中进行调度这个中间过程可能都要花费几秒钟，所以如果每个 Task 都非常快就跑完了，就会在 Task 开始和结束的时候浪费太多的时间。

配置 Task 的 JVM 重用可以改善该问题，参数是 mapred.job.reuse.jvm.num.tasks，默认值是 1，表示一个 JVM 上最多可以顺序执行的 Task 数目（属于同一个 Job）是 1，也就是说一个 Task 启动一个 JVM。这个值可以在 mapred-site.xml 中进行更改，当设置成多个就意味着多个 Task 运行在同一个 JVM 上，但不是同时执行，是排队顺序执行。

② 如果输入的文件非常大，如 1 TB 大小，可以考虑将 HDFS 上的每个 blocksize 设大，如设成 256 MB 或者 512 MB。

4. ReduceTask 的并行度

ReduceTask 的并行度同样影响着整个 Job 的执行并发度和执行效率，但与 MapTask 的并发数由切片数决定不同，ReduceTask 数量是可以直接手动设置的，相应的参数为 job.setNumReduceTasks(number)，默认值是 1，如果手动设置为 4，则表示运行 4 个 ReduceTask，设置为 0，则表示不运行 ReduceTask，也就是没有 Reduce 阶段，只有 Map 阶段。如果数据分布不均匀，就有可能在 Reduce 阶段产生数据倾斜。

5. ReduceTask 并行度的决定机制

ReduceTask 的并行度受以下几个参数影响：

- job.setNumReduceTasks(number);
- job.setReducerClass(MyReducer.class);
- job.setPartitionerClass(MyPTN.class)。

我们分以下几种情况讨论参数带来的并行度影响。

① 如果 number 为 1，并且 job.setReducerClass 已经设置为自定义 Reducer，ReduceTask 的个数就是 1，不管用户编写的 MapReduce 程序有没有设置 Partitioner，该分区组件都不会起作用。

② 如果 number 没有设置，并且 job.setReducerClass 已经设置为自定义 Reducer，ReduceTask 的个数就是 1。在默认的分区组件的影响下，用户设置的 number 只要大于 1，都是可以正常执行的。在设置自定义的分区组件时，需要注意：设置的 ReduceTask 的个数必须等于分区编号中的最大值＋1，最好的情况下，分区编号都是连续的，那么 ReduceTask 的个数＝分区编号的总个数＝分区编号中的最大值＋1。

③ 如果 number≥2，并且 job.setReducerClass 已经设置为自定义 Reducer，ReduceTask 的个数就是 number。

④ 如果设置了 number 的个数，但是没有设置自定义的 Reducer，那么该 MapReduce 程序不代表没有 Reduce 阶段。真正的 Reducer 中的逻辑就是调用父类 Reducer 中的默认实现逻辑：原样输出 ReduceTask 的个数，也就是 number。

⑤ 如果一个 MapReduce 程序中不想有 Reduce 阶段，那么只需要设置 job.setNumberReudceTasks(0)即可，如此，整个 MapReduce 程序只有 Map 阶段，没有 Reduce 阶段。

3.3　MapReduce 的多 Job 串联和全局计数器

3.3.1　MapReduce 的多 Job 串联

1. 需求分析

一个稍复杂的数据处理逻辑往往需要多个 MapReduce 程序串联处理，多 Job 的串联可以借助于 MapReduce 框架的 JobControl 实现。

2. 实例

以下有两个 MapReduce 任务，分别是 Flow 的 SumMR 和 SortMR，其中有依赖关系：SumMR 的输出是 SortMR 的输入。所以 SortMR 的启动是在 SumMR 完成之后。MapReduce 程序代码参考如下：

```java
Configuration conf1 = new Configuration();
Configuration conf2 = new Configuration();
Job job1 = Job.getInstance(conf1);
Job job2 = Job.getInstance(conf2);
job1.setJarByClass(MRScore3.class);
job1.setMapperClass(MRMapper3_1.class);
//job.setReducerClass(ScoreReducer3.class);
job1.setMapOutputKeyClass(IntWritable.class);
job1.setMapOutputValueClass(StudentBean.class);
job1.setOutputKeyClass(IntWritable.class);
job1.setOutputValueClass(StudentBean.class);
job1.setPartitionerClass(CoursePartitioner2.class);
job1.setNumReduceTasks(4);
Path inputPath = new Path("/home/hadoop/input");
Path outputPath = new Path("/home/hadoop/output_01");
FileInputFormat.setInputPaths(job1, inputPath);
FileOutputFormat.setOutputPath(job1, outputPath);
job2.setMapperClass(MRMapper3_2.class);
job2.setReducerClass(MRReducer3_2.class);
job2.setMapOutputKeyClass(IntWritable.class);
job2.setMapOutputValueClass(StudentBean.class);
job2.setOutputKeyClass(StudentBean.class);
job2.setOutputValueClass(NullWritable.class);
Path inputPath2 = new Path("/home/hadoop/output_01");
Path outputPath2 = new Path("/home/hadoop/output_01_end");
FileInputFormat.setInputPaths(job2, inputPath2);
FileOutputFormat.setOutputPath(job2, outputPath2);
JobControl control = new JobControl("Score3");
ControlledJob aJob = new ControlledJob(job1.getConfiguration());
ControlledJob bJob = new ControlledJob(job2.getConfiguration());
// 设置作业依赖关系
bJob.addDependingJob(aJob);
control.addJob(aJob);
control.addJob(bJob);
Thread thread = new Thread(control);
thread.start();
while(! control.allFinished()) {
    thread.sleep(1000);
}
System.exit(0);
```

3.3.2 全局计数器

MapReduce 计数器是什么？计数器（Counter）是用来记录 Job 的执行进度和状态的，它的作用可以理解为日志。我们可以在程序的某个位置插入计数器，记录数据或者进度的变化情况。

  MapReduce 计数器能做什么？MapReduce 计数器为我们提供一个窗口，用于观察 MapReduce Job 运行期的各种细节数据，对 MapReduce 性能调优很有帮助，MapReduce 性能优化的评估大部分都是基于这些 Counter 的数值表现出来的。

  MapReduce 都有哪些内置计数器？MapReduce 自带了许多默认 Counter，现在我们来分析这些默认 Counter 的含义，方便大家观察 Job 结果，如输入的字节数、输出的字节数、Map 端输入/输出的字节数和条数、Reduce 端输入/输出的字节数和条数等。

1. 任务计数器

  在任务执行过程中，任务计数器采集任务的相关信息，每个作业的所有任务的结果都会被聚集起来。例如，MAP_INPUT_RECORDS 计数器统计每个 Map 任务输入记录的总数，并在一个作业的所有 Map 任务上进行聚集，使得最终数字是整个作业的所有输入记录的总数。任务计数器由其关联任务维护，并定期发送给 TaskTracker，再由 TaskTracker 发送给 JobTracker。因此，计数器能够被全局地聚集。下面分别介绍各种任务计数器。

  (1) MapReduce 任务计数器

  MapReduce 任务计数器的 groupName 为 org.apache.hadoop.mapreduce.TaskCounter，它包含的计数器如表 3-1 所示。

表 3-1   MapReduce 任务计数器

计数器名称	说明
Map 输入的记录数（MAP_INPUT_RECORDS）	作业中所有 Map 已处理的输入记录数。每当 RecorderReader 读到一条记录并将其传给 Map 的 map() 函数时，该计数器的值增加
Map 跳过的记录数（MAP_SKIPPED_RECORDS）	作业中所有 Map 跳过的输入记录数
Map 输入的字节数（MAP_INPUT_BYTES）	作业中所有 Map 已处理的未经压缩的输入数据的字节数。每当 RecorderReader 读到一条记录并将其传给 Map 的 map() 函数时，该计数器的值增加
分片 Split 的原始字节数（SPLIT_RAW_BYTES）	由 Map 读取的输入-分片对象的字节数。这些对象描述分片元数据（文件的位移和长度），而不是分片数据自身，因此总规模是小的
Map 输出的记录数（MAP_OUTPUT_RECORDS）	作业中所有 Map 产生的输出记录数。每当某一个 Map 的 Context 调用 write() 方法时，该计数器的值增加
Map 输出的字节数（MAP_OUTPUT_BYTES）	作业中所有 Map 产生的未经压缩的输出数据的字节数。每当某一个 Map 的 Context 调用 write() 方法时，该计数器的值增加
Map 输出的物化字节数（MAP_OUTPUT_MATERIALIZED_BYTES）	Map 输出后实写到磁盘上的字节数。若 Map 输出压缩功能被启用，则会在该计数器值上反映出来
Combine 输入的记录数（COMBINE_INPUT_RECORDS）	作业中所有 Combiner（如果有）已处理的输入记录数。Combiner 的迭代器每读一个值，该计数器的值增加
Combine 输出的记录数（COMBINE_OUTPUT_RECORDS）	作业中所有 Combiner（如果有）已产生的输出记录数。每当一个 Combiner 的 Context 调用 write() 方法时，该计数器的值增加
Reduce 输入的组（REDUCE_INPUT_GROUPS）	作业中所有 Reducer 已经处理的不同码分组的个数。每当某一个 Reducer 的 reduce() 方法被调用时，该计数器的值增加

续表

计数器名称	说明
Reduce 输入的记录数（REDUCE_INPUT_RECORDS）	作业中所有 Reducer 已经处理的输入记录数。每当某个 Reducer 的迭代器读一个值时，该计数器的值增加。如果所有 Reducer 已经处理完所有输入，则该计数器的值与计数器"Map 输出的记录数"的值相同
Reduce 输出的记录数（REDUCE_OUTPUT_RECORDS）	作业中所有 Reducer 已经产生的输出记录数。每当某一个 Reducer 的 Context 调用 write()方法时，该计数器的值增加
Reduce 跳过的组数（REDUCE_SKIPPED_GROUPS）	作业中所有 Reducer 已经跳过的不同码分组的个数
Reduce 跳过的记录数（REDUCE_SKIPPED_RECORDS）	作业中所有 Reducer 已经跳过的输入记录数
Reduce 经过 Shuffle 的字节数（REDUCE_SHUFFLE_BYTES）	Shuffle 将 Map 的输出数据复制到 Reducer 中的字节数
溢出的记录数（SPILLED_RECORDS）	作业中所有 Map 和 Reduce 任务溢出到磁盘的记录数
CPU 毫秒（CPU_MILLISECONDS）	总计的 CPU 时间，以毫秒为单位，由/proc/cpuinfo 获取
物理内存字节数（PHYSICAL_MEMORY_BYTES）	一个任务所用物理内存的字节数，由/proc/cpuinfo 获取
虚拟内存字节数（VIRTUAL_MEMORY_BYTES）	一个任务所用虚拟内存的字节数，由/proc/cpuinfo 获取
有效的堆字节数（COMMITTED_HEAP_BYTES）	在 JVM 中的总有效内存量（以字节为单位），可由 Runtime().getRuntime().totaoMemory()获取
GC 运行时间毫秒数（GC_TIME_MILLIS）	在任务执行过程中，垃圾收集器（GC，Garbage Collector）花费的时间（以毫秒为单位），可由 GarbageCollectorMXBean.getCollectionTime()获取；该计数器并未出现在 1.x 版本中
由 Shuffle 传输的 Map 输出数（SHUFFLED_MAPS）	由 Shuffle 传输到 Reducer 的 Map 输出文件数
失败的 Shuffle 数（SHUFFLE_MAPS）	在 Shuffle 过程中，发生复制错误的 Map 输出文件数，该计数器并没有出现在 1.x 版本中
被合并的 Map 输出数	在 Shuffle 过程中，在 Reduce 端被合并的 Map 输出文件数，该计数器没有出现在 1.x 版本中

（2）文件系统计数器

文件系统计数器的 groupName 为 org.apache.hadoop.mapreduce.FileSystemCounter，它包含的计数器如表 3-2 所示。

表 3-2　文件系统计数器

计数器名称	说明
文件系统的读字节数（BYTES_READ）	由 Map 和 Reduce 等任务在各个文件系统中读取的字节数，各个文件系统分别对应一个计数器，文件系统可以是 Local、HDFS、S3 和 KFS 等
文件系统的写字节数（BYTES_WRITTEN）	由 Map 和 Reduce 等任务在各个文件系统中写的字节数

（3）FileInputFormat 计数器

FileInputFormat 计数器的 groupName 为 org.apache.hadoop.mapreduce.lib.input.FileInputFormatCounter，它包含的计数器如表 3-3 所示。

表 3-3　FileInputFormat 计数器

计数器名称	说明
读取的字节数（BYTES_READ）	由 Map 任务通过 FileInputFormat 读取的字节数

（4）FileOutputFormat 计数器

FileOutputFormat 计数器的 groupName 为 org.apache.hadoop.mapreduce.lib.input.FileOutputFormatCounter，它包含的计数器如表 3-4 所示。

表 3-4　FileOutputFormat 计数器

计数器名称	说明
写的字节数（BYTES_WRITTEN）	由 Map 任务（针对仅含 Map 的作业）或者 Reduce 任务通过 FileOutputFormat 写的字节数

2. 作业计数器

作业计数器由 JobTracker（或者 YARN）维护，因此无须在网络间传输数据，这一点与包括"用户定义的计数器"在内的其他计数器不同。这些计数器都是作业级别的统计量，其值不会随着任务运行而改变。作业计数器的 groupName 为 org.apache.hadoop.mapreduce.JobCounter，它包含的计数器如表 3-5 所示。

表 3-5　作业计数器

计数器名称	说明
启用的 Map 任务数（TOTAL_LAUNCHED_MAPS）	启动的 Map 任务数，包括以推测执行方式启动的任务
启用的 Reduce 任务数（TOTAL_LAUNCHED_REDUCES）	启动的 Reduce 任务数，包括以推测执行方式启动的任务
失败的 Map 任务数（NUM_FAILED_MAPS）	失败的 Map 任务数
失败的 Reduce 任务数（NUM_FAILED_REDUCES）	失败的 Reduce 任务数
数据本地化的 Map 任务数（DATA_LOCAL_MAPS）	与输入数据在同一节点上的 Map 任务数
机架本地化的 Map 任务数（RACK_LOCAL_MAPS）	与输入数据在同一机架范围内，但不在同一节点上的 Map 任务数
其他本地化的 Map 任务数（OTHER_LOCAL_MAPS）	与输入数据不在同一机架范围内的 Map 任务数。由于机架之间的宽带资源相对较少，Hadoop 会尽量让 Map 任务靠近输入数据执行，因此该计数器的值一般比较小
Map 任务的总运行时间（SLOTS_MILLIS_MAPS）	Map 任务的总运行时间，单位为毫秒。该计数器包括以推测执行方式启动的任务
Reduce 任务的总运行时间（SLOTS_MILLIS_REDUCES）	Reduce 任务的总运行时间，单位为毫秒。该计数器包括以推测执行方式启动的任务

续表

计数器名称	说明
在保留槽之后，Map 任务等待的总时间（FALLOW_SLOTS_MILLIS_MAPS）	在为 Map 任务保留槽之后所花费的总等待时间，单位为毫秒
在保留槽之后，Reduce 任务等待的总时间（FALLOW_SLOTS_MILLIS_REDUCES）	在为 Reduce 任务保留槽之后所花费的总等待时间，单位为毫秒

3.3.3 计数器该如何使用？

下面我们来介绍如何使用计数器。

1. 定义计数器

（1）枚举声明计数器

```
// 自定义枚举变量 Enum
Counter counter = context.getCounter(Enum enum)
```

（2）自定义计数器

```
// 自己命名 groupName 和 counterName
Counter counter = context.getCounter(String groupName,String counterName)
```

2. 为计数器赋值

（1）初始化计数器

```
counter.setValue(long value);// 设置初始值
```

（2）计数器自增

```
counter.increment(long incr);// 增加计数
```

3. 获取计数器的值

（1）获取枚举计数器的值

```
Configuration conf = new Configuration();
Job job = new Job(conf,"MyCounter");
job.waitForCompletion(true);
Counters counters = job.getCounters();
Counter counter = counters.findCounter(LOG_PROCESSOR_COUNTER.BAD_RECORDS_LONG);
//查找枚举计数器，假设 Enum 变量为 BAD_RECORDS_LONG
long value=counter.getValue();//获取计数值
```

（2）获取自定义计数器的值

```
Configuration conf = new Configuration();
Job job = new Job(conf,"MyCounter");
job.waitForCompletion(true);
Counters counters = job.getCounters();
```

```
Counter counter = counters.findCounter("ErrorCounter","toolong");
//假设 groupName 为 ErrorCounter,counterName 为 toolong
long value = counter.getValue();// 获取计数值
```

（3）获取内置计数器的值

```
Configuration conf = new Configuration();
Job job = new Job(conf, "MyCounter");
job.waitForCompletion(true);
Counters counters = job.getCounters();
// 查找作业运行启动的 Reduce 个数的计数器,groupName 和 counterName 可以从内置计数器表格中查
询(前面已经列举)
Counter counter = counters.findCounter("org.apache.hadoop.mapreduce.JobCounter","TOTAL_LAUNCHED_REDUCES");
// 假设 groupName 为 org.apache.hadoop.mapreduce.JobCounter,counterName 为 TOTAL_LAUNCHED_REDUCES
long value = counter.getValue();// 获取计数值
```

（4）获取所有计数器的值

```
Configuration conf = new Configuration();
Job job = new Job(conf, "MyCounter");
Counters counters = job.getCounters();
for (CounterGroup group : counters) {
  for (Counter counter : group) {
    System.out.println(counter.getDisplayName() + ":" + counter.getName() + ":" +
                counter.getValue());
  }
}
```

3.3.4 MapReduce 框架 Partitioner 分区

1. Partitioner 分区类的作用是什么？

在进行 MapReduce 计算时,有时候需要把最终的输出数据分到不同的文件中。例如:按照省份划分的话,需要把同一省份的数据放到一个文件中;按照性别划分的话,需要把同一性别的数据放到一个文件中。我们知道最终的输出数据来自 Reduce 任务,那么,如果要得到多个文件,就要有同样数量的 Reduce 任务在运行。Reduce 任务的数据来自 Map 任务,也就是说 Map 任务要划分数据,将不同的数据分配给不同的 Reduce 任务运行。Map 任务划分数据的过程就称作 Partition,负责实现划分数据的类称作 Partitioner。

Partitioner 类的源码如下：

```
package org.apache.hadoop.mapreduce.lib.partition;
import org.apache.hadoop.mapreduce.Partitioner;
/** Partition keys by their {@link Object#hashCode()}. */
public class HashPartitioner<K, V> extends Partitioner<K, V> {
  /** Use {@link Object#hashCode()} to partition. */
```

```
    public int getPartition(K key, V value, int numReduceTasks) {
      //默认使用 key 的 hash 值与 int 的最大值,避免出现数据溢出的情况
      return (key.hashCode() & Integer.MAX_VALUE) % numReduceTasks;
    }
}
```

2. getPartition( )的 3 个参数

HashPartitioner 是处理 Map 任务输出的,getPartition( )方法有 3 个形参,源码中 key、value 分别指的是 Map 任务的输出,numReduceTasks 指的是设置的 Reduce 任务数量,默认值是 1。任何整数与 1 相除的余数肯定是 0,也就是说,getPartition( )方法的返回值默认总是 0,即 Map 任务的输出总是送给一个 Reduce 任务,最终只能输出到一个文件中。

据此分析,如果想要最终输出到多个文件中,在 Map 任务中应将数据划分到多个区中。那么,我们只需要按照一定的规则让 getPartition( )方法的返回值是 0,1,2,3,…即可。大部分情况下,我们都会使用默认的分区函数,但有时我们又有一些特殊的需求,而需要定制 Partition 来完成我们的业务。

3. 案例分析

案例:按照能否被 5 除尽进行分区。如果除以 5 的余数是 0,则放在 0 号分区;如果除以 5 的余数不是 0,则放在 1 号分区。参考代码如下:

```java
import org.apache.hadoop.io.IntWritable;
import org.apache.hadoop.mapreduce.Partitioner;
public class FivePartitioner extends Partitioner<IntWritable, IntWritable>{
    /**
     * 我们的需求:按照能否被 5 除尽进行分区
     *
     * 1. 如果除以 5 的余数是 0,则放在 0 号分区
     * 2. 如果除以 5 的余数不是 0,则放在 1 号分区
     */
    @Override
    public int getPartition(IntWritable key, IntWritable value, int numPartitions) {
        int intValue = key.get();
        if(intValue % 5 == 0){
            return 0;
        }else{
            if(intValue % 2 == 0){
                return 1;
            }else{
                return 2;
            }
        }
    }
}
```

在运行 MapReduce 程序时,只需在主函数里加入如下代码即可:

```java
job.setPartitionerClass(FivePartitioner.class);
job.setNumReduceTasks(3);//设置为 3
```

3.3.5 MapReduce 框架 Combiner 分区

1. 对 Combiner 分区的理解

Combiner 其实属于优化方案,在实际场景中,由于带宽限制,应该尽量减少 Map 和 Reduce 之间的数据传输数量。Combiner 在 Map 端把 key 相同的键值对合并在一起并计算,计算规则与 Reduce 一致,所以 Combiner 也可以看作特殊的 Reducer。

执行 Combine 操作要求开发者必须在程序中设置了 Combiner〔程序中通过 job.setCombinerClass(myCombine.class)自定义 Combine 操作〕。Combiner 组件是用来做局部汇总的,就在 MapTask 中进行汇总;Reducer 组件是用来做全局汇总的,最终进行一次汇总。

2. 在哪里使用 Combiner?

① Map 输出数据根据分区排序完成后,在写入文件之前会执行一次 Combine 操作,前提是作业中设置了这个操作。

② 如果 Map 输出比较大,溢出文件个数大于 3(此值可以通过属性 min.num.spills.for.combine 配置),在合并的过程中还会执行 Combine 操作。

不是每种作业都可以做 Combine 操作,只有满足以下条件才可以:

① Combiner 只能对一个 MapTask 的中间结果进行汇总。

② 如果想使用 Reducer 直接充当 Combiner,那么必须满足:Reducer 的输入和输出 key/value 类型是一致的。

③ 处于两个不同节点的 MapTask 的结果不能 Combine 到一起。

④ 处于同一个节点的两个 MapTask 的结果不能 Combine 到一起。

⑤ 求最大值、求最小值、求和、去重时可直接使用 Reducer 充当 Combiner,但是求平均值时不能直接使用 Reducer 充当 Combiner。

3.4 YARN 的资源调度

1. YARN 功能介绍

YARN(Yet Another Resource Negotiator)是一个资源调度平台,负责为运算程序提供服务器运算资源,相当于一个分布式的操作系统平台,而 MapReduce 等运算程序则相当于运行于操作系统之上的应用程序。

YARN 是 Hadoop 2.x 版本中的一个新特性,它的出现其实是为了解决第一代 MapReduce 编程框架的不足,提高集群环境下的资源利用率,这些资源包括内存、磁盘、网络、I/O 等。Hadoop 2.x 版本中重新设计的 YARN 集群具有更好的扩展性、可用性、可靠性、向后兼容性,以及能支持除 MapReduce 以外的更多分布式运算程序。

YARN 并不清楚用户提交的程序的运行机制,YARN 只提供运算资源的调度,YARN 中的主管角色叫作 ResourceManager,具体提供运算资源的角色叫作 NodeManager。这样一来,YARN 其实就与运行的用户程序完全解耦,这意味着 YARN 上可以运行各种类型的分布式运算程序(MapReduce 只是其中的一种),如 MapReduce 程序、Storm 程序、Spark 程序、Tez

程序……所以 Spark、Storm 等运算框架都可以整合在 YARN 上运行,只要它们各自的框架中有符合 YARN 规范的资源请求机制即可。YARN 就成为一个通用的资源调度平台,从此企业中以前存在的各种运算集群都可以整合在一个物理集群上,提高资源利用率,方便数据共享。

2. 原 MapReduce 框架的不足

① JobTracker 是集群事务的集中处理点,存在单点故障。

② JobTracker 需要完成的任务太多,既要维护 Job 的状态,又要维护 Job 的 Task 的状态,造成过多的资源消耗。

③ 在 TaskTracker 端,用 MapTask/ReduceTask 作为资源的表示过于简单,没有考虑 CPU、内存等资源情况,当把两个需要消耗大内存的 Task 调度到一起,很容易出现内存溢出(OOM)。

④ 把资源强制划分为 MapSlot/ReduceSlot,当只有 MapTask 时,ReduceSlot 不能用,当只有 ReduceTask 时,MapSlot 不能用,容易造成资源利用不足。

3. 新版 YARN 架构的优点

YARN/MRv2 最基本的想法是将原 JobTracker 主要的资源管理和 Job 调度/监视功能分开,作为两个单独的守护进程。有一个全局的 ResourceManager(简称 RM)且每个 Application 有一个 ApplicationMaster(简称 AM),Application 相当于 MapReduce Job 或者 DAG Job。ResourceManager 和 NodeManager(简称 NM)组成了基本的数据计算框架。RM 协调集群的资源利用,任何 Client 或者运行着的 AM 想要运行 Job 或者 Task,都得向 RM 申请一定的资源。AM 是一个框架特殊的库,MapReduce 框架有它自己的 AM 实现,用户也可以实现自己的 AM,在运行的时候,AM 会与 NM 一起来启动和监视 Task。

4. YARN 的重要概念

(1) ResourceManager

ResourceManager 是基于应用程序对集群资源需求进行调度的 YARN 集群主控节点,负责协调和管理整个集群(所有 NodeManager)的资源,响应用户提交的不同类型应用程序的解析、调度、监控等工作。ResourceManager 会为每一个 Application 启动一个 MRAppMaster,并且 MRAppMaster 分散在各个 NodeManager 节点上。MRAppMaster 进程由 ApplicationMaster 实现,使得 MapReduce 可以直接运行在 YARN 上,用于管理作业的生命周期。

ResourceManager 主要由两个组件构成:调度器(Scheduler)和应用程序管理器(ApplicationsManager,ASM)。YARN 集群的主节点 ResourceManager 的职责包括:

- 处理客户端请求;
- 启动或监控 MRAppMaster;
- 监控 NodeManager;
- 资源的分配与调度。

(2) NodeManager

NodeManager 是 YARN 集群中真正资源的提供者,是真正执行应用程序的容器的提供者,监控应用程序的资源(CPU、内存、硬盘、网络)使用情况,并通过心跳向集群资源调度器

ResourceManager 进行汇报以更新自己的健康状态。同时其也会监督 Container 的生命周期管理,监控每个 Container 的资源(内存、CPU 等)使用情况,追踪节点健康状况,管理日志和不同应用程序用到的附属服务(auxiliary service)。

YARN 集群的从节点 NodeManager 的职责包括:
- 管理单个节点上的资源;
- 处理来自 ResourceManager 的命令;
- 处理来自 MRAppMaster 的命令。

(3) MRAppMaster

MRAppMaster 对应一个应用程序,职责是向资源调度器申请执行任务的资源容器、执行任务、监控整个任务的执行、跟踪整个任务的状态、处理任务失败以及异常情况。

(4) Container

Container 是一个抽象出来的逻辑资源单位。Container 是 ResourceManager Scheduler 服务动态分配的资源构成,它包括该节点上的一定数量的 CPU、内存、磁盘、网络等信息,MapReduce 程序的所有 Task 都是在一个 Container 里完成执行的,Container 的大小是可以动态调整的。

(5) ASM

ASM 负责管理整个系统中的所有应用程序,职责包括应用程序提交、与调度器协商资源以启动 MRAppMaster、监控 MRAppMaster 的运行状态并在失败时重新启动它等。

(6) Scheduler

Scheduler 根据应用程序的资源需求进行资源分配,不参与应用程序具体的执行和监控等工作。资源分配的单位就是 Container,调度器是一个可插拔的组件,用户可以根据自己的需求实现自己的调度器。YARN 本身为我们提供了多种直接可用的调度器,如 FIFO、Fair Scheduler 和 Capacity Scheduler 等。

5. YARN 作业执行流程

① 用户向 YARN 提交应用程序,其中包括 MRAppMaster 程序、启动 MRAppMaster 的命令、用户程序等。

② ResourceManager 为该程序分配第一个 Container,并与对应的 NodeManager 通信,要求它在这个 Container 中启动应用程序 MRAppMaster。

③ MRAppMaster 首先向 ResourceManager 注册,这样用户可以直接通过 ResourceManager 查看应用程序的运行状态,然后 MRAppMaster 将为各个任务申请资源,并监控任务的运行状态,直到运行结束,重复步骤④~⑦。

④ MRAppMaster 采用轮询的方式通过 RPC 协议向 ResourceManager 申请和领取资源。

⑤ MRAppMaster 申请到资源后,便与对应的 NodeManager 通信,要求它启动任务。

⑥ NodeManager 为任务设置好运行环境(包括环境变量、jar 包、二进制程序等)后,将任务启动命令写到一个脚本中,并通过运行该脚本启动任务。

⑦ 各个任务通过某个 RPC 协议向 MRAppMaster 汇报自己的状态和进度,以便 MRAppMaster 随时掌握各个任务的运行状态,从而可以在任务失败的时候重新启动任务。

⑧ 应用程序运行完成后,MRAppMaster 向 ResourceManager 注销并关闭自己。

本章小结

① Hadoop MapReduce 是一个使用简易的软件框架，基于它写出来的应用程序能够运行在由上千个商用机器组成的大型集群上，并以一种可靠容错的方式并行处理太字节级别的数据集。一个 Map/Reduce 作业（Job）通常会把输入的数据集切分为若干独立的数据块，由 Map 任务（Task）以完全并行的方式处理它们。框架会先对 Map 的输出进行排序，然后把结果输入给 Reduce 任务。通常作业的输入和输出都会被存储在文件系统中。整个框架负责任务的调度和监控，以及重新执行已经失败的任务。

通常，MapReduce 框架和分布式文件系统是运行在一组相同的节点上的，也就是说，计算节点和存储节点通常在一起。这种配置允许框架在那些已经存好数据的节点上高效地调度任务，这可以使整个集群的网络带宽被非常高效地利用。

MapReduce 框架由一个单独的 Master JobTracker 和每个集群节点的一个 Slave TaskTracker 共同组成。Master 负责调度构成一个作业的所有任务，这些任务分布在不同的 Slave 节点上。Master 节点监控它们的执行，重新执行已经失败的任务，而 Slave 节点仅负责执行由 Master 节点指派的任务。

应用程序至少应该指明输入/输出的位置（路径），并通过实现合适的接口或抽象类来提供 map() 和 reduce() 函数。再加上其他作业的参数，就构成了作业配置（Job Configuration）。然后，Hadoop 的 Job-Client 提交作业（jar 包/可执行程序等）和配置信息给 JobTracker，后者负责分发这些作业和配置信息给 Slave、调度任务并监控它们的执行，同时提供状态和诊断信息给 Job-Client。

② Apache YARN 从整体上还是属于 Master/Slave 模型，主要依赖于 3 个组件来实现功能。第一个组件是 ResourceManager，是集群资源的仲裁者，它包括两部分：一是可插拔式的 Scheduler，一是 ApplicationsManager，用于管理集群中的用户作业。第二个组件是每个节点上的 NodeManager，管理该节点上的用户作业和工作流，也会不断发送自己的 Container 使用情况给 ResourceManager。第三个组件是 ApplicationMaster，用户作业生命周期的管理者，它的主要功能就是向 ResourceManager（全局的）申请计算资源（Container），并且和 NodeManager 交互来执行和监控具体的 Task。

第 4 章
HBase 分布式数据库

自 1970 年以来,关系数据库被用于存储数据和维护有关问题的解决方案。大数据出现后,很多企业实现大数据处理并从中受益,开始选择 Hadoop 大数据平台这样的解决方案。Hadoop 使用分布式文件系统来存储大数据,并使用 MapReduce 来处理。Hadoop 擅长存储各种格式的庞大的数据,甚至可以处理非结构化的数据。

4.1 HBase 数据库概述

4.1.1 HBase 数据库的使用场景

1. Hadoop 的限制

Hadoop 只能执行批量处理,并且只以顺序方式访问数据。这意味着必须搜索整个数据集,即使是最简单的搜索工作。当处理结果在另一个庞大的数据集中,也是按顺序处理一个庞大的数据集。在这一点上需要一个新的解决方案,需要能够访问数据中的任何点(随机访问)。与 HDFS 一样,HBase 存储目标主要依靠横向扩展,通过不断增加廉价的商用服务器,来提高计算和存储能力。

Hadoop HBase 组件属于 NoSQL 数据库的一种。NoSQL 泛指非关系型的数据库,随着互联网 Web 2.0 网站的兴起,传统的关系数据库在处理 Web 2.0 网站,特别是超大规模和高并发的社交网络服务(SNS)类型的 Web 2.0 纯动态网站时,已经显得力不从心,出现了很多难以克服的问题,而非关系型的数据库则由于其特点得到了非常迅速的发展。NoSQL 数据库的产生就是为了解决大规模数据集合多重数据种类带来的挑战,特别是大数据应用难题。常见的 NoSQL 数据库(如 HBase、Cassandra、CouchDB、Dynamo 和 MongoDB)都是一些存储大量数据和以随机方式访问数据的数据库。

综上所述,NoSQL 数据库是在如下背景下产生的。

① 海量数据存储成为瓶颈,单台机器无法负载大量数据。

② 单台机器读写请求成为存储海量数据时高并发、大规模请求的瓶颈。

③ 随着数据规模越来越大,大量业务场景开始考虑数据存储横向扩展,使得存储服务可以增加/删除,而关系数据库更专注于单台机器。

2. 数据类型

在当前生产环境下,数据主要包括如下几种类型。

① 结构化数据:数据结构字段含义确定、清晰,典型的如数据库中的表结构。

② 半结构化数据:具有一定结构,但语义不够确定,典型的如 HTML 网页,有些字段是确定的(title),有些字段不确定(table)。

③ 非结构化数据:杂乱无章的数据,很难按照一个概念进行抽取,无规律性。

3. HBase 的定义

HBase 是一个构建在 HDFS 上的分布式列存储系统,HBase 是基于 Google BigTable 模型开发的典型 key/value 系统。HBase 是 Apache Hadoop 生态系统中的重要一员,主要用于海量结构化数据存储。从逻辑上讲,HBase 将数据按照表、行和列进行存储,与 Hadoop 集群一样,HBase 数据库存储目标主要依靠横向扩展,通过不断增加廉价的商用服务器,来提高计算和存储能力。HBase 是建立在 HDFS 之上,用来提供高可靠性、高性能、列存储、可伸缩、多版本的 NoSQL 的分布式数据存储系统,实现对大型数据的实时、随机读写访问。

4. HBase 中表的特点

HBase 中表的特点如图 4-1 所示。

- 海量存储:一张表可以有上十亿行,上百万列。
- 列式存储:面向列(族)的存储和权限控制,列(族)独立检索。
- 稀疏:为空(NULL)的列并不占用存储空间,因此,表可以设计得非常稀疏。
- 高并发:并发写入数据,不支持文件的随机修改。
- 极易扩展:HBase 运行在 Hadoop 集群上,可扩展到上千台服务器。

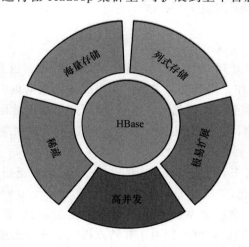

图 4-1 HBase 中表的特点

5. HBase 表结构逻辑视图

(1) Rowkey

Rowkey 的概念和 MySQL 中的主键是完全一样的,HBase 使用 Rowkey 来唯一区分某一行的数据。Rowkey 对 HBase 性能的影响非常大,因此 Rowkey 的设计显得尤为重要。设计的时候要兼顾基于 Rowkey 的单行查询,也要键入 Rowkey 的范围扫描。Rowkey 可以是任意字符串(最大长度是 64 KB,实际应用中长度一般为 10~100 B,最好的值是 16 B)。在 HBase 内部 Rowkey 保存为字节数组,HBase 会对表中的数据按照 Rowkey 进行字典排序。

(2) Column

HBase 列可理解成 MySQL 列，属于某一个 ColumnFamily(列族)，格式为 familyName：columnName，每条记录可动态添加。

(3) ColumnFamily

列族是 HBase 引入的概念，HBase 通过列族划分数据的存储，列族可以包含任意多的列，实现灵活的数据存取。就像是家族的概念，我们知道一个家族是由很多个家庭组成的，列族也类似，列族是由任意多的列组成的。HBase 表在创建的时候就必须指定列族，就像关系数据库在创建的时候必须指定具体的列一样。HBase 的列族不是越多越好，官方推荐的是列族最好小于或者等于 3，我们使用的场景一般是 1 个列族。

(4) TimeStamp

TimeStamp(时间戳)对 HBase 来说至关重要，因为它是实现 HBase 多版本的关键。在 HBase 中使用不同的 TimeStamp 来标识相同 Rowkey 行对应的不同版本的数据。HBase 中通过 Rowkey 和 Column 确定的为一个存储单元，称为 Cell。每个 Cell 都保存着同一份数据的多个版本。版本通过时间戳来索引，时间戳的类型是 64 位整型。时间戳可以由 HBase(在数据写入时自动)赋值，此时时间戳是精确到毫秒的当前系统时间。时间戳也可以由客户显式赋值。如果应用程序要避免数据版本冲突，就必须自己生成具有唯一性的时间戳。每个 Cell 中，不同版本的数据按照时间倒序排列，即最新的数据排在最前面。

(5) Cell

Cell 中的数据是没有类型的，全部以字节码形式存储，可以存在多个不同版本的值。

4.1.2　HBase 数据库的安装

通过浏览器打开 HBase 的官网 https://hbase.apache.org，可以看到有图 4-2 所示的选择。

图 4-2　HBase 的官网下载

此处我们的 Hadoop 选择的版本是 2.7.5，HBase 数据库选择的版本是 1.2.6，官方的版本说明可通过网址 http://hbase.apache.org/downloads.html 查看。

1. 下载安装包

下载 HBase 安装包 hbase-1.2.6-bin.tar.gz，这里给大家提供一个下载的镜像地址 http://mirrors.hust.edu.cn/apache/hbase/，可以加快国内镜像的下载速度。

2. 上传服务器并解压缩到指定目录

```
[hadoop@hadoop1 ~]$ ls
hbase-1.2.6-bin.tar.gz
[hadoop@hadoop1 ~]$ tar -zxvf hbase-1.2.6-bin.tar.gz -C apps/
```

3. 修改配置文件

配置文件目录在安装包的 conf 文件夹中。

（1）修改 hbase-env.sh 文件

```
[hadoop@hadoop1 conf]$ vi hbase-env.sh
export JAVA_HOME=/usr/local/jdk1.8.0_73
export HBASE_MANAGES_ZK=false
```

（2）修改 hbase-site.xml 文件

```
[hadoop@hadoop1 conf]$ vi hbase-site.xml
<configuration>
    <property>
        <!-- 指定 HBase 在 HDFS 上存储的路径 -->
        <name>hbase.rootdir</name>
        <value>hdfs://hadoop1:9000/hbase</value>
    </property>
    <property>
        <!-- 指定 HBase 是分布式的 -->
        <name>hbase.cluster.distributed</name>
        <value>true</value>
    </property>
    <property>
        <!-- 指定 ZooKeeper 的地址，多个用","分割 -->
        <name>hbase.zookeeper.quorum</name>
        <value>hadoop1:2181,hadoop2:2181,hadoop3:2181,hadoop4:2181</value>
    </property>
</configuration>
```

（3）修改 regionservers 文件

```
[hadoop@hadoop1 conf]$ vi regionservers
hadoop1
hadoop2
hadoop3
hadoop4
```

（4）修改 backup-masters 文件

该文件是不存在的，请自行创建，目的是指定 Master 的备份：

```
[hadoop@hadoop1 conf]$ vi backup-masters
hadoop4
```

(5) 修改 hdfs-site.xml 和 core-site.xml 文件

最重要的一步是要把 Hadoop 的 hdfs-site.xml 和 core-site.xml 放到 hbase-1.2.6/conf 下：

```
[hadoop@hadoop1 conf]$ cd ~/apps/hadoop-2.7.5/etc/hadoop/
[hadoop@hadoop1 hadoop]$ cp core-site.xml hdfs-site.xml ~/apps/hbase-1.2.6/conf/
```

4. 将 HBase 安装包分发到其他节点

分发之前先删除 hbase 目录下的 docs 文件夹，再分发到集群中其他的服务器：

```
[Hadoop@Hadoop1 hbase-1.2.6]$ rm -rf docs/
[hadoop@hadoop1 apps]$ scp -r hbase-1.2.6/ hadoop2:$PWD
[hadoop@hadoop1 apps]$ scp -r hbase-1.2.6/ hadoop3:$PWD
[hadoop@hadoop1 apps]$ scp -r hbase-1.2.6/ hadoop4:$PWD
```

5. 配置环境变量

所有服务器都要进行配置：

```
[hadoop@hadoop1 apps]$ vi ~/.bashrc
export HBASE_HOME=/home/hadoop/apps/hbase-1.2.6
export PATH=$PATH:$HBASE_HOME/bin
```

使环境变量立即生效：

```
[hadoop@hadoop1 apps]$ source ~/.bashrc
```

6. 启动 HBase 集群

严格按照启动顺序进行。

(1) 启动 ZooKeeper 集群

HBase 数据库软件内置了简化的 ZooKeeper 功能，在实际应用场景中，我们建议单独配置 ZooKeeper 集群，实现 HBase 集群的功能。每个 ZooKeeper 节点都要执行以下命令：

```
[hadoop@hadoop1 apps]$ zkServer.sh start
ZooKeeper JMX enabled by default
Using config: /home/hadoop/apps/zookeeper-3.4.10/bin/../conf/zoo.cfg
Starting zookeeper ... STARTED
```

(2) 启动 HDFS 集群及 YARN 集群

如果需要运行 MapReduce 程序则启动 YARN 集群，否则不需要启动。

```
[hadoop@hadoop1 apps]$ start-dfs.sh
Starting namenodes on [hadoop1 hadoop2]
hadoop2: starting namenode, logging to /home/hadoop/apps/hadoop-2.7.5/logs/hadoop-hadoop-namenode-hadoop2.out
hadoop1: starting namenode, logging to /home/Hadoop/apps/Hadoop-2.7.5/logs/hadoop-hadoop-namenode-hadoop1.out
```

```
    hadoop3: starting datanode, logging to /home/hadoop/apps/hadoop-2.7.5/logs/hadoop-hadoop-
datanode-hadoop3.out
    hadoop4: starting datanode, logging to /home/hadoop/apps/hadoop-2.7.5/logs/hadoop-hadoop-
datanode-hadoop4.out
    hadoop2: starting datanode, logging to /home/hadoop/apps/hadoop-2.7.5/logs/hadoop-hadoop-
datanode-hadoop2.out
    hadoop1: starting datanode, logging to /home/hadoop/apps/hadoop-2.7.5/logs/hadoop-hadoop-
datanode-hadoop1.out
    Starting journal nodes [hadoop1 hadoop2 hadoop3]
    hadoop3: starting journalnode, logging to /home/hadoop/apps/hadoop-2.7.5/logs/hadoop-hadoop-
journalnode-hadoop3.out
    hadoop2: starting journalnode, logging to /home/Hadoop/apps/hadoop-2.7.5/logs/hadoop-hadoop-
journalnode-hadoop2.out
    hadoop1: starting journalnode, logging to /home/Hadoop/apps/hadoop-2.7.5/logs/hadoop-hadoop-
journalnode-hadoop1.out
    Starting ZK Failover Controllers on NN hosts [hadoop1 hadoop2]
    hadoop2: starting zkfc, logging to /home/hadoop/apps/hadoop-2.7.5/logs/hadoop-hadoop-zkfc-
hadoop2.out
    hadoop1: starting zkfc, logging to /home/hadoop/apps/hadoop-2.7.5/logs/hadoop-hadoop-zkfc-
hadoop1.out
```

启动完成之后检查以下 NameNode 的状态：

```
[hadoop@hadoop1 apps]$ hdfs haadmin -getServiceState nn1
standby
[hadoop@hadoop1 apps]$ hdfs haadmin -getServiceState nn2
active
```

(3) 启动 HBase 集群

在保证 ZooKeeper 集群和 HDFS 集群启动正常的情况下，启动 HBase 集群的命令为"start-hbase.sh"，在哪一个节点上执行启动命令，哪一个节点就是主节点。

```
[hadoop@hadoop1 conf]$ start-hbase.sh
    starting master, logging to /home/hadoop/apps/hbase-1.2.6/logs/hbase-hadoop-master-hadoop1.out
    Java HotSpot(TM) 64-Bit Server VM warning: ignoring option PermSize=128m; support was removed in
8.0
    Java HotSpot(TM) 64-Bit Server VM warning: ignoring option MaxPermSize=128m; support was removed
in 8.0
    hadoop3: starting regionserver, logging to /home/hadoop/apps/hbase-1.2.6/logs/hbase-hadoop-
regionserver-hadoop3.out
    hadoop4: starting regionserver, logging to /home/Hadoop/apps/hbase-1.2.6/logs/hbase-hadoop-
regionserver-hadoop4.out
    hadoop2: starting regionserver, logging to /home/Hadoop/apps/hbase-1.2.6/logs/hbase-hadoop-
regionserver-hadoop2.out
    hadoop3: Java HotSpot(TM) 64-Bit Server VM warning: ignoring option PermSize=128m; support was
removed in 8.0
    hadoop3: Java HotSpot(TM) 64-Bit Server VM warning: ignoring option MaxPermSize=128m; support
was removed in 8.0
```

```
    hadoop4: Java HotSpot(TM) 64-Bit Server VM warning: ignoring option PermSize = 128m; support was
removed in 8.0
    hadoop4: Java HotSpot(TM) 64-Bit Server VM warning: ignoring option MaxPermSize = 128m; support
was removed in 8.0
    hadoop2: Java HotSpot(TM) 64-Bit Server VM warning: ignoring option PermSize = 128m; support was
removed in 8.0
    hadoop2: Java HotSpot(TM) 64-Bit Server VM warning: ignoring option MaxPermSize = 128m; support
was removed in 8.0
    hadoop1: starting regionserver, logging to /home/Hadoop/apps/hbase-1.2.6/logs/hbase-hadoop-
regionserver-hadoop1.out
    hadoop4: starting master, logging to /home/hadoop/apps/hbase-1.2.6/logs/hbase-hadoop-master-
hadoop4.out
```

观察启动日志可以看到：
① 首先在命令执行节点上启动 Master。
② 然后分别在 hadoop2、hadoop3、hadoop4 上启动 RegionServer。
③ 然后在 backup-masters 文件中配置的备用节点上再启动一个 Master 主进程。

7．验证数据库进程是否启动正常

① 检查各进程是否启动正常，使用 jps 命令，我们大致可以看到如下进程：

```
4960 HMaster
2960 QuorumPeerMain
3169 NameNode
3699 DFSZKFailoverController
3285 DataNode
5098 HRegionServer
5471 Jps
3487 JournalNode
```

② 通过访问浏览器页面，我们输入 URL 地址"http://主机名或 IP 地址:16010"，如图 4-3 所示。

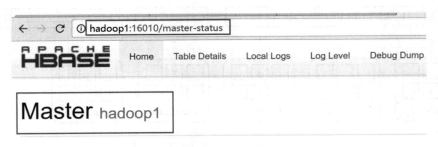

图 4-3　HBase 集群 Web 访问界面

③ 手动启动服务。如果有节点相应的进程没有启动，那么可以手动启动，可以使用如下

命令来启动 HMaster 进程：

```
[hadoop@hadoop3 conf]$ hbase-daemon.sh start master
starting master, logging to /home/hadoop/apps/hbase-1.2.6/logs/hbase-Hadoop-master-hadoop3.out
Java HotSpot(TM) 64-Bit Server VM warning: ignoring option PermSize=128m; support was removed in 8.0
Java HotSpot(TM) 64-Bit Server VM warning: ignoring option MaxPermSize=128m; support was removed in 8.0
```

启动 HRegionServer 进程：

```
[hadoop@hadoop3 conf]$ hbase-daemon.sh start regionserver
```

4.2 HBase 数据库物理架构

4.2.1 HBase 集群节点类型

HBase 数据库集群整体架构如图 4-4 所示。

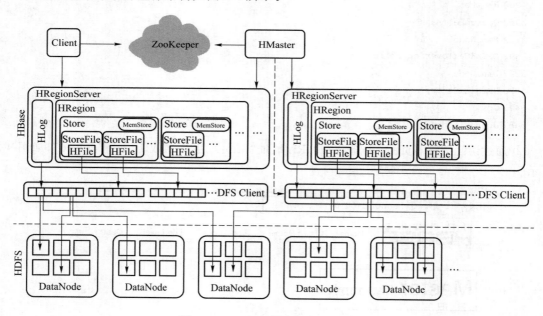

图 4-4　HBase 数据库集群整体架构

1. ZooKeeper 服务器

要保证任何时候集群中只有一个 Running Master，Master 与 RegionServers 启动时会向 ZooKeeper 注册。默认情况下，HBase 管理 ZooKeeper 实例，如启动或者停止 ZooKeeper，ZooKeeper 的引入使得 Master 不再出现单点故障。ZooKeeper 主要用于如下场景：

- 存储所有 Region 的寻址入口。
- 实时监控 RegionServer 的状态，将 RegionServer 的上线和下线信息实时通知给 Master。

- 存储 HBase 的 Schema 和 Table 元数据。

2. Master 服务器

Master 服务器的主要功能总结如下：

- 管理用户对 Table 的增删改查操作。
- 在 RegionSplit 后，分配新 Region。
- 负责 RegionServer 的负载均衡，调整 Region 分布。
- 在 RegionServer 停机后，负责失效 RegionServer 上 Region 的重新分配。
- HMaster 失效仅会导致所有元数据无法修改，表达数据读写可以正常运行。

3. RegionServer 服务器

RegionServer 服务器主要实现如下功能：

- RegionServer 维护 Region，处理对这些 Region 的 I/O 请求。
- RegionServer 负责切分在运行过程中变得过大的 Region。
- Client 访问 HBase 上数据的过程并不需要 Master 参与，寻址先访问 ZooKeeper 再访问 RegionServer，数据读写访问 RegionServer。
- HRegionServer 主要负责响应用户 I/O 请求，向 HDFS 中读写数据，是 HBase 中最核心的模块。

4. Client 客户端

Client 客户端主要实现如下功能：

- 整合 HBase 集群的入口。
- 使用 HBase RPC 机制与 HMaster 和 HRegionServer 通信。
- 与 HMaster 通信，进行管理类操作。
- 与 HRegionServer 通信，进行读写类操作。
- 包含访问 HBase 的接口，Client 维护着一些 Cache 来加快对 HBase 的访问，如访问 Region 的位置信息。

4.2.2 HBase 数据存储

1. Region

Region 是分布式数据库 HBase 存储的最小单元，但不是数据存储最小的单元。Region 由一个或多个 Store 组成，每个 Store 保存一个 ColumnFamily。每个 Store 又由一个 MemStore 和 0 个或多个 StoreFile 组成，MemStore 存储在内存中，StoreFile 存储在 HDFS 上。Region 是 HBase 中分布式存储和负载均衡的最小单元，不同的 Region 分布到不同的 RegionServer 上，但 Region 不会拆分到不同的 RegionServer 上。

一张 HBase 表会按照行划分为若干个 Region，每一个 Region 被分配给一台特定的 RegionServer。每一个 Region 内部还要依据列族划分为若干个 Store，每个 Store 中的数据会落地到若干个 HFile 文件中。Region 体积会随着数据插入而不断增长，到一定阈值后会分裂，随着 Region 的分裂，一台 RegionServer 管理的 Region 会越来越多。HMaster 会根据 RegionServer 管理的 Region 数做负载均衡。Region 中的数据拥有一个内存缓存——MemStore，数据的访问优先在 MemStore 中进行。因为空间有限，所以 MemStore 中的数据需要定期 Flush 到 StoreFile 中，每次 Flush 都是生成新的 StoreFile。StoreFile 的数量随着时间会不断增加，RegionServer 会定期将大量 StoreFile 进行合并（Merge）。Region 分布架构可以参考图 4-5。

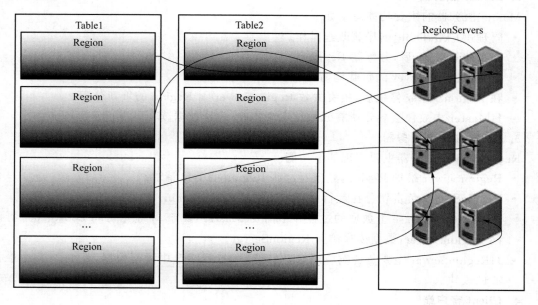

图 4-5　Region 分布架构

2. StoreFile

StoreFile 存储结构如图 4-6 所示。

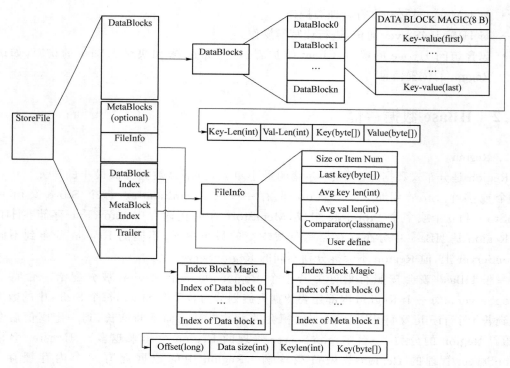

图 4-6　StoreFile 存储结构

具体字段说明如下。

- DataBlocks 段：保存表中的数据，这部分可以被压缩。
- MetaBlocks 段（可选的）：保存用户自定义的 KV 对，可以被压缩。
- FileInfo 段：HFile 的元信息，不被压缩，用户也可以在这一部分添加自己的元信息。
- DataBlockIndex 段：DataBlock 的索引。每条索引的 key 是被索引的 Block 的第一条记录的 key。
- MetaBlockIndex 段（可选的）：MetaBlock 的索引。
- Trailer：这一段是定长的，保存了每一段的偏移量，读取一个 HFile 时，会首先读取 Trailer，然后 DataBlockIndex 会被读取到内存中。这样在检索某个 key 时，不需要扫描整个 HFile，而只需从内存中找到 key 所在的 Block，通过一次磁盘 I/O 将整个 Block 读取到内存中，再找到需要的 key，DataBlockIndex 采用最近最少使用（LRU）机制淘汰。
- HFile 的 DataBlock、MetaBlock 通常采用压缩方式存储，压缩之后可以大大减少网络 I/O 和磁盘 I/O，随之而来的当然是需要花费 CPU 进行压缩和解压缩。

4.3 HBase 数据库操作

4.3.1 HBase 命令行的启动

1. 进入 HBase 命令行

在集群中的服务器节点上，执行命令"hbase shell"，会进入 HBase Shell 客户端：

[hadoop@hadoop1 ~]$ hbase shell

```
SLF4J: Class path contains multiple SLF4J bindings.
SLF4J: Found binding in [jar:file:/home/hadoop/apps/hbase-1.2.6/lib/slf4j-log4j12-1.7.5.jar!/org/slf4j/impl/StaticLoggerBinder.class]
SLF4J: Found binding in [jar:file:/home/hadoop/apps/hadoop-2.7.5/share/hadoop/common/lib/slf4j-log4j12-1.7.10.jar!/org/slf4j/impl/StaticLoggerBinder.class]
SLF4J: See http://www.slf4j.org/codes.html#multiple_bindings for an explanation.
SLF4J: Actual binding is of type [org.slf4j.impl.Log4jLoggerFactory]
HBase Shell; enter 'help<RETURN>' for list of supported commands.
Type "exit<RETURN>" to leave the HBase Shell
Version 1.2.6, r239b80456118175b340b2e562a5568b5c744252e, Mon May 29 02:25:32 CDT 2017
hbase(main):001:0>
```

2. HBase 命令行的帮助

大家可以查看一下提示，其实这是一个 help 帮助信息，讲述了怎么获得帮助，怎么退出客户端：

```
HBase Shell; enter 'help<RETURN>' for list of supported commands.
Type "exit<RETURN>" to leave the HBase Shell
```

help 获取帮助
    help:获取所有命令提示
    help "dml":获取一组命令的提示
    help "put":获取一个单独命令的提示
exit 退出 hbase shell 客户端

4.3.2 HBase 表的操作

表的操作包括创建表(create)、查看表列表(list)、查看表的详细信息(desc)、删除表(drop)、清空表(truncate)和修改表的定义(alter)。

1. 创建表(create)

可以输入以下命令查看帮助命令:

hbase(main):001:0> help 'create'

```
Create table; pass table name, a dictionary of specifications per
column family, and optionally a dictionary of table configuration.
Dictionaries are described below in the GENERAL NOTES section.
Examples:
  hbase> create 't1', {NAME => 'f1', VERSIONS => 5}
  hbase> create 't1', {NAME => 'f1'}, {NAME => 'f2'}, {NAME => 'f3'}
  hbase> # The above in shorthand would be the following:
  hbase> create 't1', 'f1', 'f2', 'f3'
  hbase> create 't1', {NAME => 'f1', VERSIONS => 1, TTL => 2592000, BLOCKCACHE => true}
  hbase> create 't1', 'f1', {SPLITS => ['10', '20', '30', '40']}
  hbase> create 't1', 'f1', {SPLITS_FILE => 'splits.txt'}
  hbase> # Optionally pre-split the table into NUMREGIONS, using
  hbase> # SPLITALGO ("HexStringSplit", "UniformSplit" or classname)
  hbase> create 't1', 'f1', {NUMREGIONS => 15, SPLITALGO => 'HexStringSplit'}
```

可以看到其中一条提示:

hbase> create 't1', {NAME => 'f1'}, {NAME => 'f2'}, {NAME => 'f3'}

其中,t1 是表名,f1,f2,f3 是列族的名,如:

```
hbase(main):002:0> create 'myHbase',{NAME => 'myCard',VERSIONS => 5}
0 row(s) in 3.1270 seconds
=> Hbase::Table - myHbase
hbase(main):003:0>
```

创建了一个名为 myHbase 的表,表里面有 1 个列族,名为 myCard,保留 5 个版本信息。

2. 查看表列表(list)

可以输入以下命令查看帮助命令:

hbase(main):003:0> help 'list'

```
List all tables in hbase. Optional regular expression parameter could
be used to filter the output. Examples:
    hbase> list
    hbase> list 'abc.*'
    hbase> list 'ns:abc.*'
    hbase> list 'ns:.*'
hbase(main):004:0>
```

直接输入 list 进行查看：

```
hbase(main):004:0> list
TABLE
myHbase
1 row(s) in 0.0650 seconds
=> ["myHbase"]
hbase(main):005:0>
```

只有一条结果，就是刚刚创建的表 myHbase。

3. 查看表的详细信息（desc）

一对大括号就相当于一个列族。

```
hbase(main):006:0> desc 'myHbase'
Table myHbase is ENABLED
myHbase
COLUMN FAMILIES DESCRIPTION
{NAME =>'myCard', BLOOMFILTER =>'ROW', VERSIONS =>'5', IN_MEMORY =>'false', KEEP_DELETED_CELLS =>'FALSE', DATA_BLOCK_ENCODING =>'NONE', TTL =>'FOREVER', COMPRESSION =>'NONE', MIN_VERSIONS =>'0', BLOCKCACHE =>'true', BLOCKSIZE =>'65536', REPLICATION_SCOPE =>'0'}
1 row(s) in 0.2160 seconds
hbase(main):007:0>
```

4. 修改表的定义（alter）

（1）添加一个列族

```
hbase(main):007:0> alter 'myHbase', NAME =>'myInfo'
Updating all regions with the new schema...
1/1 regions updated.
Done.
0 row(s) in 2.0690 seconds
hbase(main):008:0> desc 'myHbase'
Table myHbase is ENABLED
myHbase
COLUMN FAMILIES DESCRIPTION
{NAME =>'myCard', BLOOMFILTER =>'ROW', VERSIONS =>'5', IN_MEMORY =>'false', KEEP_DELETED_CELLS =>'FALSE', DATA_BLOCK_ENCODING =>'NONE', TTL =>'FOREVER', COMPRESSION =>'NONE', MIN_VERSIONS =>'0', BLOCKCACHE =>'true', BLOCKSIZE =>'65536', REPLICATION_SCOPE =>'0'}
```

```
{NAME =>'myInfo', BLOOMFILTER =>'ROW', VERSIONS =>'1', IN_MEMORY =>'false', KEEP_DELETED_
CELLS =>'FALSE', DATA_BLOCK_ENCODING =>'NONE', TTL =>'FOREVER', COMPRESSION =>'NONE', MIN_
VERSIONS =>'0', BLOCKCACHE =>'true', BLOCKSIZE =>'65536', REPLICATION_SCOPE =>'0'}
 2 row(s) in 0.0420 seconds
hbase(main):009:0>
```

(2) 删除一个列族

```
hbase(main):009:0> alter 'myHbase', NAME =>'myCard', METHOD =>'delete'
Updating all regions with the new schema...
1/1 regions updated.
Done.
0 row(s) in 2.1920 seconds
hbase(main):010:0> desc 'myHbase'
Table myHbase is ENABLED
myHbase
COLUMN FAMILIES DESCRIPTION
{NAME =>'myInfo', BLOOMFILTER =>'ROW', VERSIONS =>'1', IN_MEMORY =>'false', KEEP_DELETED_
CELLS =>'FALSE', DATA_BLOCK_ENCODING =>'NONE', TTL =>'FOREVER', COMPRESSION =>'NONE', MIN_
VERSIONS =>'0', BLOCKCACHE =>'true', BLOCKSIZE =>'65536', REPLICATION_SCOPE =>'0'}
1 row(s) in 0.0290 seconds
hbase(main):011:0>
```

要删除一个列族也可以执行如下命令：

```
hbase(main):011:0> alter 'myHbase', 'delete' =>'myCard'
```

要添加列族 myTest，同时删除列族 myInfo，可以尝试使用如下命令：

```
hbase(main):011:0> alter 'myHbase', {NAME =>'myTest'}, {NAME =>'myInfo', METHOD =>'delete'}
Updating all regions with the new schema...
1/1 regions updated.
Done.
Updating all regions with the new schema...
1/1 regions updated.
Done.
0 row(s) in 3.8260 seconds
hbase(main):012:0> desc 'myHbase'
Table myHbase is ENABLED
myHbase
COLUMN FAMILIES DESCRIPTION
{NAME =>'myTest', BLOOMFILTER =>'ROW', VERSIONS =>'1', IN_MEMORY =>'false', KEEP_DELETED_
CELLS =>'FALSE', DATA_BLOCK_ENCODING =>'NONE', TTL =>'FOREVER', COMPRESSION =>'NONE', MIN_
VERSIONS =>'0', BLOCKCACHE =>'true', BLOCKSIZE =>'65536', REPLICATION_SCOPE =>'0'}
1 row(s) in 0.0410 seconds
hbase(main):013:0>
```

5. 清空表（truncate）

```
hbase(main):013:0> truncate 'myHbase'
Truncating 'myHbase' table (it may take a while):
- Disabling table...
- Truncating table...
0 row(s) in 3.6760 seconds
hbase(main):014:0>
```

6. 删除表（drop）

```
hbase(main):014:0> drop 'myHbase'
ERROR: Table myHbase is enabled. Disable it first.
Here is some help for this command:
Drop the named table. Table must first be disabled:
  hbase> drop 't1'
  hbase> drop 'ns1:t1'
hbase(main):015:0>
```

直接删除表会报错，根据提示需要先停用表，我们执行如下操作：

```
hbase(main):015:0> disable 'myHbase'
0 row(s) in 2.2620 seconds
hbase(main):016:0> drop 'myHbase'
0 row(s) in 1.2970 seconds
hbase(main):017:0> list
TABLE
0 row(s) in 0.0110 seconds
=> []
hbase(main):018:0>
```

4.3.3　HBase 表中数据的操作

我们可以创建 user_info 表，包含 base_info、extra_info 两个列族，执行如下命令：

```
hbase(main):018:0> create 'user_info',{NAME =>'base_info',VERSIONS => 3 },{NAME =>'extra_info', VERSIONS => 1 }
0 row(s) in 4.2670 seconds
=> Hbase::Table - user_info
hbase(main):019:0>
```

1. 录入数据

查看帮助，需要传入表名、Rowkey、列族名、值等，注意数据格式的顺序。

```
hbase(main):019:0> help 'put'
```

```
Put a cell 'value' at specified table/row/column and optionally
timestamp coordinates.  To put a cell value into table 'ns1:t1' or 't1'
at row 'r1' under column 'c1' marked with the time 'ts1', do:
  hbase> put 'ns1:t1', 'r1', 'c1', 'value'
  hbase> put 't1', 'r1', 'c1', 'value'
  hbase> put 't1', 'r1', 'c1', 'value', ts1
  hbase> put 't1', 'r1', 'c1', 'value', {ATTRIBUTES =>{'mykey'=>'myvalue'}}
  hbase> put 't1', 'r1', 'c1', 'value', ts1, {ATTRIBUTES =>{'mykey'=>'myvalue'}}
  hbase> put 't1', 'r1', 'c1', 'value', ts1, {VISIBILITY =>'PRIVATE|SECRET'}
The same commands also can be run on a table reference. Suppose you had a reference
t to table 't1', the corresponding command would be:
  hbase> t.put 'r1', 'c1', 'value', ts1, {ATTRIBUTES =>{'mykey'=>'myvalue'}}
hbase(main):020:0>
```

向 user_info 表中插入信息:

```
hbase(main):020:0> put 'user_info', 'user0001', 'base_info:name', 'zhangsan1'
hbase(main):021:0> put 'user_info', 'rk0001', 'base_info:name', 'zhangsan'
```

此处可以多添加几条数据。

2. 查询数据

获取 user_info 表中 Rowkey 为 user0001 的所有信息:

```
hbase(main):022:0> get 'user_info', 'user0001'
COLUMN                  CELL
base_info:name          timestamp = 1522320801670, value = zhangsan1
1 row(s) in 0.1310 seconds
hbase(main):023:0>
```

获取 user_info 表中 Rowkey 为 rk0001 的 base_info 列族的所有信息:

```
hbase(main):025:0> get 'user_info', 'rk0001', 'base_info'
COLUMN                  CELL
base_info:name          timestamp = 1522321247732, value = zhangsan
1 row(s) in 0.0320 seconds
hbase(main):026:0>
```

查询 user_info 表中的所有信息:

```
hbase(main):026:0> scan 'user_info'
ROW                     COLUMN + CELL
rk0001                  column = base_info:name, timestamp = 1522321247732, value = zhangsan
user0001                column = base_info:name, timestamp = 1522320801670, value = zhangsan1
2 row(s) in 0.0970 seconds
hbase(main):027:0>
```

查询 user_info 表中列族为 base_info 的信息:

```
hbase(main):027:0> scan 'user_info', {COLUMNS =>'base_info'}
```

```
ROW                          COLUMN + CELL
rk0001                       column = base_info:name, timestamp = 1522321247732, value = zhangsan
user0001                     column = base_info:name, timestamp = 1522320801670, value = zhangsan1
2 row(s) in 0.0620 seconds
hbase(main):028:0>
```

3．删除数据

删除 user_info 表中 Rowkey 为 rk0001，列标示符为 base_info:name 的数据：

```
hbase(main):028:0> delete 'user_info', 'rk0001', 'base_info:name'
0 row(s) in 0.0780 seconds
hbase(main):029:0> scan 'user_info', {COLUMNS => 'base_info'}
ROW                          COLUMN + CELL
user0001                     column = base_info:name, timestamp = 1522320801670, value = zhangsan1
1 row(s) in 0.0530 seconds
hbase(main):030:0>
```

4.4　HBase 数据库的 API 操作

HBase 具有 Java 语言编写的增删改查操作，并具有 Java 原生 API，因此它提供了编程访问数据操纵语言（DML）。HBaseConfiguration 类添加 HBase 的配置到配置文件中，这个类属于 org.apache.hadoop.hbase 包。

HBase 的 API 操作如下：

```java
import java.io.IOException;
import java.util.Date;
import org.apache.hadoop.conf.Configuration;
import org.apache.hadoop.hbase.HBaseConfiguration;
import org.apache.hadoop.hbase.HColumnDescriptor;
import org.apache.hadoop.hbase.HTableDescriptor;
import org.apache.hadoop.hbase.TableName;
import org.apache.hadoop.hbase.client.Admin;
import org.apache.hadoop.hbase.client.Connection;
import org.apache.hadoop.hbase.client.ConnectionFactory;
import org.apache.hadoop.hbase.client.Delete;
import org.apache.hadoop.hbase.client.Get;
import org.apache.hadoop.hbase.client.Put;
import org.apache.hadoop.hbase.client.Result;
import org.apache.hadoop.hbase.client.ResultScanner;
import org.apache.hadoop.hbase.client.Scan;
import org.apache.hadoop.hbase.client.Table;
import com.study.hbase.service.HBaseUtils;
```

```java
public class HBaseUtilsImpl implements HBaseUtils {
    private static final String ZK_CONNECT_KEY = "hbase.zookeeper.quorum";
    private static final String ZK_CONNECT_VALUE = "hadoop1:2181,hadoop2:2181,hadoop3:2181";
    private static Connection conn = null;
    private static Admin admin = null;
    public static void main(String[] args) throws Exception {
        getConnection();
        getAdmin();
        HBaseUtilsImpl hbu = new HBaseUtilsImpl();
        //hbu.getAllTables();
        //hbu.descTable("people");
        //String[] infos = {"info","family"};
        //hbu.createTable("people", infos);
        //String[] add = {"cs1","cs2"};
        //String[] remove = {"cf1","cf2"};
        //HColumnDescriptor hc = new HColumnDescriptor("sixsixsix");
        //hbu.modifyTable("stu",hc);
        //hbu.getAllTables();
        hbu.putData("huoying", "rk001", "cs2", "name", "aobama",new Date().getTime());
        hbu.getAllTables();
        conn.close();
    }
    // 获取连接
    public static Connection getConnection() {
        // 创建一个可以用来管理 HBase 配置信息的 conf 对象
        Configuration conf = HBaseConfiguration.create();
        // 设置当前的程序寻找的 HBase 在哪里
        conf.set(ZK_CONNECT_KEY, ZK_CONNECT_VALUE);
        try {
            conn = ConnectionFactory.createConnection(conf);
        } catch (IOException e) {
            e.printStackTrace();
        }
        return conn;
    }
    // 获取管理员对象
    public static Admin getAdmin() {
        try {
            admin = conn.getAdmin();
        } catch (IOException e) {
            e.printStackTrace();
        }
        return admin;
    }
    // 查询所有表
    @Override
```

```java
    public void getAllTables() throws Exception {
        //获取列族的描述信息
        HTableDescriptor[] listTables = admin.listTables();
        for (HTableDescriptor listTable : listTables) {
            //转化为表名
            String tbName = listTable.getNameAsString();
            //获取列的描述信息
            HColumnDescriptor[] columnFamilies = listTable.getColumnFamilies();
            System.out.println("tableName:" + tbName);
            for(HColumnDescriptor columnFamilie : columnFamilies) {
                //获取列族的名字
                String columnFamilyName = columnFamilie.getNameAsString();
                System.out.print("\t" + "columnFamilyName:" + columnFamilyName);
            }
            System.out.println();
        }
    }
    // 创建表,传参:表名和列族的名字
    @Override
    public void createTable(String tableName, String[] family) throws Exception {
        TableName name = TableName.valueOf(tableName);
        //判断表是否存在
        if(admin.tableExists(name)) {
            System.out.println("table 已经存在!");
        }else {
            //表的列族示例
            HTableDescriptor htd = new HTableDescriptor(name);
            //向列族中添加列的信息
            for(String str : family) {
                HColumnDescriptor hcd = new HColumnDescriptor(str);
                htd.addFamily(hcd);
            }
            //创建表
            admin.createTable(htd);
            //判断表是否创建成功
            if(admin.tableExists(name)) {
                System.out.println("table 创建成功");
            }else {
                System.out.println("table 创建失败");
            }
        }
    }
    // 创建表,传参:封装好的多个列族
    @Override
    public void createTable(HTableDescriptor htds) throws Exception {
        //获得表的名字
```

```java
        String tbName = htds.getNameAsString();
        admin.createTable(htds);
}
// 创建表,传参:表名和封装好的多个列族
@Override
public void createTable(String tableName, HTableDescriptor htds) throws Exception {
    TableName name = TableName.valueOf(tableName);
    if(admin.tableExists(name)) {
        System.out.println("table 已经存在!");
    }else {
        admin.createTable(htds);
        boolean flag = admin.tableExists(name);
        System.out.println(flag ? "创建成功" : "创建失败");
    }
}
// 查看表的列族属性
@Override
public void descTable(String tableName) throws Exception {
    //转化为表名
    TableName name = TableName.valueOf(tableName);
    //判断表是否存在
    if(admin.tableExists(name)) {
        //获取表中列族的描述信息
        HTableDescriptor tableDescriptor = admin.getTableDescriptor(name);
        //获取列族中列的信息
        HColumnDescriptor[] columnFamilies = tableDescriptor.getColumnFamilies();
        for(HColumnDescriptor columnFamily : columnFamilies) {
            System.out.println(columnFamily);
        }
    }else {
        System.out.println("table 不存在");
    }
}
// 判断表是否存在
@Override
public boolean existTable(String tableName) throws Exception {
    TableName name = TableName.valueOf(tableName);
    return admin.tableExists(name);
}
// disable 表
@Override
public void disableTable(String tableName) throws Exception {
    TableName name = TableName.valueOf(tableName);
    if(admin.tableExists(name)) {
        if(admin.isTableEnabled(name)) {
            admin.disableTable(name);
```

```java
        }else {
            System.out.println("table 不是活动状态");
        }
    }else {
        System.out.println("table 不存在");
    }
}
// drop 表
@Override
public void dropTable(String tableName) throws Exception {
    //转化为表名
    TableName name = TableName.valueOf(tableName);
    //判断表是否存在
    if(admin.tableExists(name)) {
        //判断表是否处于可用状态
        boolean tableEnabled = admin.isTableEnabled(name);
        if(tableEnabled) {
            //使表处于不可用状态
            admin.disableTable(name);
        }
        //删除表
        admin.deleteTable(name);
        //判断表是否存在
        if(admin.tableExists(name)) {
            System.out.println("删除失败");
        }else {
            System.out.println("删除成功");
        }
    }else {
        System.out.println("table 不存在");
    }
}
// 修改表(增加和删除)
@Override
public void modifyTable(String tableName) throws Exception {
    //转化为表名
    TableName name = TableName.valueOf(tableName);
    //判断表是否存在
    if(admin.tableExists(name)) {
        //判断表是否处于可用状态
        boolean tableEnabled = admin.isTableEnabled(name);
        if(tableEnabled) {
            //使表处于不可用状态
            admin.disableTable(name);
        }
        //根据表名得到表
```

```java
            HTableDescriptor tableDescriptor = admin.getTableDescriptor(name);
            //创建列族结构对象
            HColumnDescriptor columnFamily1 = new HColumnDescriptor("cf1".getBytes());
            HColumnDescriptor columnFamily2 = new HColumnDescriptor("cf2".getBytes());
            tableDescriptor.addFamily(columnFamily1);
            tableDescriptor.addFamily(columnFamily2);
            //替换该表所有的列族
            admin.modifyTable(name, tableDescriptor);
        }else {
            System.out.println("table 不存在");
        }
    }
    // 修改表(增加和删除)
    @Override
    public void modifyTable(String tableName, String[] addColumn, String[] removeColumn) throws Exception {
        //转化为表名
        TableName name = TableName.valueOf(tableName);
        //判断表是否存在
        if(admin.tableExists(name)) {
            //判断表是否处于可用状态
            boolean tableEnabled = admin.isTableEnabled(name);
            if(tableEnabled) {
                //使表处于不可用状态
                admin.disableTable(name);
            }
            //根据表名得到表
            HTableDescriptor tableDescriptor = admin.getTableDescriptor(name);
            //创建列族结构对象,添加列
            for(String add : addColumn) {
                HColumnDescriptor addColumnDescriptor = new HColumnDescriptor(add);
                tableDescriptor.addFamily(addColumnDescriptor);
            }
            //创建列族结构对象,删除列
            for(String remove : removeColumn) {
                HColumnDescriptor removeColumnDescriptor = new HColumnDescriptor(remove);
                tableDescriptor.removeFamily(removeColumnDescriptor.getName());
            }
            admin.modifyTable(name, tableDescriptor);
        }else {
            System.out.println("table 不存在");
        }
    }
    @Override
    public void modifyTable(String tableName, HColumnDescriptor hcds) throws Exception {
        //转化为表名
```

```java
        TableName name = TableName.valueOf(tableName);
        //根据表名得到表
        HTableDescriptor tableDescriptor = admin.getTableDescriptor(name);
        //获取表中所有的列族信息
        HColumnDescriptor[] columnFamilies = tableDescriptor.getColumnFamilies();
        boolean flag = false;
        //判断参数中传入的列族是否已经在表中存在
        for(HColumnDescriptor columnFamily : columnFamilies) {
            if(columnFamily.equals(hcds)) {
                flag = true;
            }
        }
        //存在提示,不存在则直接添加该列族信息
        if(flag) {
            System.out.println("该列族已经存在");
        }else {
            tableDescriptor.addFamily(hcds);
            admin.modifyTable(name, tableDescriptor);
        }
    }
    /** 添加数据
     * tableName:  表名
     * rowKey:     行键
     * familyName: 列族
     * columnName: 列名
     * value:      值
     */
    @Override
    public void putData(String tableName, String rowKey, String familyName, String columnName, String value)throws Exception {
        //转化为表名
        TableName name = TableName.valueOf(tableName);
        //添加数据之前先判断表是否存在,不存在则先创建表
        if(admin.tableExists(name)) {
        }else {
            //根据表名创建表结构
            HTableDescriptor tableDescriptor = new HTableDescriptor(name);
            //定义列族的名字
            HColumnDescriptor columnFamilyName = new HColumnDescriptor(familyName);
            tableDescriptor.addFamily(columnFamilyName);
            admin.createTable(tableDescriptor);

        }
        Table table = conn.getTable(name);
        Put put = new Put(rowKey.getBytes());
```

```java
            put.addColumn(familyName.getBytes(), columnName.getBytes(), value.getBytes());
            table.put(put);
    }
    @Override
    public void putData(String tableName, String rowKey, String familyName, String columnName, String value, long timestamp) throws Exception {
        // 转化为表名
        TableName name = TableName.valueOf(tableName);
        // 添加数据之前先判断表是否存在,不存在则先创建表
        if (admin.tableExists(name)) {
        } else {
            // 根据表名创建表结构
            HTableDescriptor tableDescriptor = new HTableDescriptor(name);
            // 定义列族的名字
            HColumnDescriptor columnFamilyName = new HColumnDescriptor(familyName);
            tableDescriptor.addFamily(columnFamilyName);
            admin.createTable(tableDescriptor);
        }
        Table table = conn.getTable(name);
        Put put = new Put(rowKey.getBytes());
        //put.addColumn(familyName.getBytes(), columnName.getBytes(), value.getBytes());
        put.addImmutable(familyName.getBytes(), columnName.getBytes(), timestamp, value.getBytes());
        table.put(put);
    }
    // 根据 Rowkey 查询数据
    @Override
    public Result getResult(String tableName, String rowKey) throws Exception {
        Result result;
        TableName name = TableName.valueOf(tableName);
        if(admin.tableExists(name)) {
            Table table = conn.getTable(name);
            Get get = new Get(rowKey.getBytes());
            result = table.get(get);
        }else {
            result = null;
        }
        return result;
    }
    // 根据 Rowkey 查询数据
    @Override
    public Result getResult(String tableName, String rowKey, String familyName) throws Exception {
        Result result;
        TableName name = TableName.valueOf(tableName);
```

```java
            if(admin.tableExists(name)) {
                Table table = conn.getTable(name);
                Get get = new Get(rowKey.getBytes());
                get.addFamily(familyName.getBytes());
                result = table.get(get);
            }else {
                result = null;
            }
            return result;
    }
    // 根据 Rowkey 查询数据
    @Override
    public Result getResult (String tableName, String rowKey, String familyName, String columnName) throws Exception {
            Result result;
            TableName name = TableName.valueOf(tableName);
            if(admin.tableExists(name)) {
                Table table = conn.getTable(name);
                Get get = new Get(rowKey.getBytes());
                get.addColumn(familyName.getBytes(), columnName.getBytes());
                result = table.get(get);
            }else {
                result = null;
            }
            return result;
    }
    // 查询指定 version
    @Override
    public Result getResultByVersion(String tableName, String rowKey, String familyName, String columnName, int versions) throws Exception {
            Result result;
            TableName name = TableName.valueOf(tableName);
            if(admin.tableExists(name)) {
                Table table = conn.getTable(name);
                Get get = new Get(rowKey.getBytes());
                get.addColumn(familyName.getBytes(), columnName.getBytes());
                get.setMaxVersions(versions);
                result = table.get(get);
            }else {
                result = null;
            }
            return result;
    }
    // scan 全表数据
    @Override
```

```java
    public ResultScanner getResultScann(String tableName) throws Exception {
        ResultScanner result;
        TableName name = TableName.valueOf(tableName);
        if(admin.tableExists(name)) {
            Table table = conn.getTable(name);
            Scan scan = new Scan();
            result = table.getScanner(scan);
        }else {
            result = null;
        }
        return result;
    }
    // scan 全表数据
    @Override
    public ResultScanner getResultScann(String tableName, Scan scan) throws Exception {
        ResultScanner result;
        TableName name = TableName.valueOf(tableName);
        if(admin.tableExists(name)) {
            Table table = conn.getTable(name);
            result = table.getScanner(scan);
        }else {
            result = null;
        }
        return result;
    }
    // 删除数据(指定的列)
    @Override
    public void deleteColumn(String tableName, String rowKey) throws Exception {
        TableName name = TableName.valueOf(tableName);
        if(admin.tableExists(name)) {
            Table table = conn.getTable(name);
            Delete delete = new Delete(rowKey.getBytes());
            table.delete(delete);
        }else {
            System.out.println("table 不存在");
        }
    }
    // 删除数据(指定的列)
    @Override
    public void deleteColumn (String tableName, String rowKey, String falilyName) throws Exception {
        TableName name = TableName.valueOf(tableName);
        if(admin.tableExists(name)) {
            Table table = conn.getTable(name);
            Delete delete = new Delete(rowKey.getBytes());
            delete.addFamily(falilyName.getBytes());
```

```
                table.delete(delete);
            }else {
                System.out.println("table 不存在");
            }
        
    }
    // 删除数据(指定的列)
    @Override
    public void deleteColumn(String tableName, String rowKey, String falilyName, String columnName) throws Exception {
        TableName name = TableName.valueOf(tableName);
        if(admin.tableExists(name)) {
            Table table = conn.getTable(name);
            Delete delete = new Delete(rowKey.getBytes());
            delete.addColumn(falilyName.getBytes(), columnName.getBytes());
            table.delete(delete);
        }else {
            System.out.println("table 不存在");
        }
    }
}
```

本 章 小 结

① HBase 是一个分布式的、面向列的开源数据库，源于 Google 的一篇论文——《BigTable：一个结构化数据的分布式存储系统》。HBase 是 Google BigTable 的开源实现，它利用 HDFS 作为其文件存储系统，利用 Hadoop MapReduce 来处理 HBase 中的海量数据，利用 ZooKeeper 作为协同服务。

② HBase 以表的形式存储数据，表整体上由行和列组成，列划分为若干个列族，列族可以包含多列。

③ HBase Master 是服务器，负责管理所有的 HRegion 服务器，HBase Master 并不存储 HBase 的任何数据，HBase 逻辑上的表可能会划分为多个 HRegion，然后存储在 HRegionServer 群中，HBase Master 中存储的是从数据到 HRegionServer 的映射。一个服务器只能运行一个 HRegion 服务器，数据的操作会记录在 HLog 中，在读取数据时，HRegion 会先访问 HMemCache 缓存，如果缓存中没有数据才会到 HStore 中查找，每一个列都会有一个 HStore 集合，每个 HStore 集合包含了很多具体的 HFile 文件，这些文件是 B 树结构的，方便快速读取。

第 5 章

Hive 数据仓库

5.1 Hive 简介

5.1.1 什么是 Hive？

Hive 由 Facebook 公司开发并开源，是基于 Apache Hadoop 的一个数据仓库工具，可以将结构化数据映射为一张数据库表并提供 HQL(Hive SQL)查询功能，底层数据存储在 HDFS 上，Hive 的本质是将 HQL 语句转换为 MapReduce 任务运行，使不熟悉 MapReduce 的用户可以很方便地利用 HQL 处理和计算 HDFS 上的结构化数据，适用于离线的批量数据计算。

数据仓库之父比尔·恩门(Bill Inmon)在 1991 年出版的 *Building the Data Warehouse*(中文译名为《建立数据仓库》)一书中所提出的定义被广泛接受——数据仓库(Data Warehouse)是一个面向主题的(subject oriented)、集成的(integrated)、相对稳定的(non-volatile)、反映历史变化的(time variant)数据集合，用于支持管理决策(decision making support)。

Hive 依赖于 HDFS 存储数据，Hive 将 HQL 转换成 MapReduce 执行，所以说 Hive 是基于 Hadoop 的一个数据仓库工具，实质就是一款基于 HDFS 的 MapReduce 计算框架，对存储在 HDFS 中的数据进行分析和管理。

直接使用 MapReduce 所面临的问题是：人员学习成本太高；项目周期要求太短；MapReduce 实现复杂查询逻辑开发难度太大。

Hive 可以给 SQL 开发工程师带来如下便利。
- 更友好的接口：操作接口采用类 SQL 的语法，提供快速开发的能力。
- 更低的学习成本：避免了写 MapReduce，减少开发人员的学习成本。
- 更好的扩展性：可自由扩展集群规模而无须重启服务，还支持用户自定义函数。

1. Hive 的特点

(1) 优点
- 可扩展性：Hive 可以自由扩展集群的规模，一般情况下不需要重启服务。
- 延展性：Hive 支持自定义函数，用户可以根据自己的需求来实现自己的函数。

- 良好的容错性：可以保障即使有节点出现问题，SQL 语句仍可完成执行。

（2）缺点
- Hive 不支持记录级别的增删改操作，但是用户可以通过查询生成新表或者将查询结果导入文件中。
- Hive 的查询延时很严重，因为 MapReduce Job 的启动过程要消耗很长时间，所以适合用在交互查询系统中。
- Hive 不支持事务，Hive 具有 SQL 数据库的外表，但应用场景完全不同，Hive 只适合用来做海量离线数据统计分析，也就是用作数据仓库。

2．Hive 的组成架构

Hive 的组成架构可以参考图 5-1。

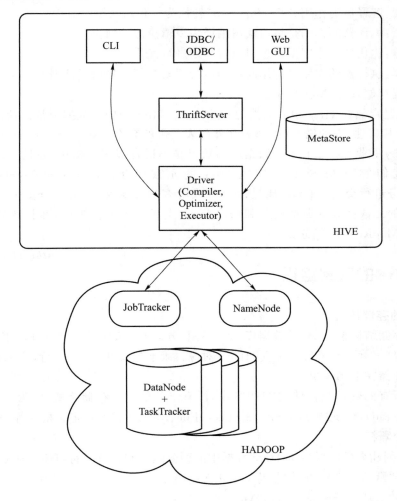

图 5-1　Hive 的组成架构

从图 5-1 中可以看出，Hive 的内部架构由以下四部分组成。

① 用户接口：CLI、JDBC/ODBC、Web GUI。
- CLI（Command Line Interface，命令行接口）采用交互形式使用 Hive 命令行与 Hive 进行交互。

- JDBC/ODBC 是 Hive 基于 JDBC(Java 数据库连接)操作提供的客户端,用户通过数据源接口连接至 Hive Server。
- Web GUI 可以通过浏览器访问 Hive。

② ThriftServer 是一种跨语言服务,让用户可以使用多种不同的语言来操纵 Hive,Thrift 是 Facebook 开发的一个软件框架,可以用来进行可扩展并且跨语言的服务的开发,Hive 集成了该服务,能让不同的编程语言调用 Hive 的接口。

③ 底层的 Driver:

Driver 组件完成 HQL 查询语句的词法分析、语法分析、编译、优化以及逻辑执行计划的生成。生成的逻辑执行计划存储在 HDFS 中,随后由 MapReduce 调用执行 Hive 的核心——驱动引擎,驱动引擎由以下四部分组成。

- 解释器:解释器的作用是将 HQL 语句转换为抽象语法树(AST)。
- 编译器:编译器的作用是将语法树编译为逻辑执行计划。
- 优化器:优化器的作用是对逻辑执行计划进行优化。
- 执行器:执行器的作用是调用底层的运行框架执行逻辑执行计划。

④ 元数据存储系统(MetaStore):

元数据就是存储在 Hive 中的数据的描述信息。Hive 中的元数据通常包括:表的名字、表的列和分区及其属性、表的属性(内部表和外部表)、表的数据所在目录。MetaStore 默认存储在自带的 Derby 数据库中。Derby 数据库的缺点是不适合多用户操作,并且数据存储目录不固定。一般我们使用 MySQL 来实现 Hive 的元数据存储,Hive 和 MySQL 之间通过 MetaStore 服务进行交互。HQL 通过命令行或者客户端提交,经过 Compiler 编译器,运用 MetaStore 中的元数据进行类型检测和语法分析,生成一个逻辑执行计划,然后通过优化处理,产生一个 MapReduce 任务。

5.1.2 Hive 的数据组织

1. Hive 的存储结构

Hive 的存储结构包括 Hive 数据库、表、视图、分区和表数据等。数据库、表、分区等都对应 HDFS 上的一个目录,表数据对应 HDFS 对应目录下的文件。

2. 数据存储在 HDFS 中

Hive 中所有的数据都存储在 HDFS 中,没有专门的数据存储格式,因为 Hive 是读模式(Schema On Read),可支持 TextFile、SequenceFile、RCFile 或者自定义格式等。

3. 数据分隔符

只需要在创建表的时候告诉 Hive 数据中的列分隔符和行分隔符,Hive 就可以解析数据。

- Hive 的默认列分隔符:控制符 Ctrl+A,\x01。
- Hive 的默认行分隔符:换行符\n。

4. Hive 中包含的数据模型

- database:在 HDFS 中表现为 ${hive.metastore.warehouse.dir} 目录下的一个文件夹。
- table:在 HDFS 中表现为所属 database 目录下的一个文件夹。
- external table:与 table 类似,不过其数据存放位置可以指定为任意 HDFS 目录路径。

- partition：在 HDFS 中表现为 table 目录下的子目录。
- bucket：在 HDFS 中表现为同一个表目录或者分区目录下根据某个字段的值进行 Hash 散列之后的多个文件。
- view：与传统数据库类似，视图是只读的，基于基本表创建。

5. Hive 的元数据

Hive 的元数据存储在关系数据库管理系统（RDBMS）中，除元数据外的其他所有数据都基于 HDFS 存储。默认情况下，Hive 元数据保存在内嵌的 Derby 数据库中，只能允许一个会话连接，只适合简单的测试，实际生产环境中不适用。为了支持多用户会话，则需要一个独立的元数据库，使用 MySQL 作为元数据库，Hive 内部对 MySQL 提供了很好的支持。

5.1.3 Hive 的表类型

1. 内部表和外部表

Hive 数据仓库中的表分为内部表和外部表两种类型，内部表和外部表均可以再分为分区表和分桶表类型。内部表和外部表的主要区别是：删除内部表，Hive 会删除表元数据和数据；删除外部表，Hive 只删除元数据，不删除数据。实际使用场景中，如何选择内部表和外部表的使用？大多数情况下，它们的区别不明显，如果数据的所有处理都在 Hive 中进行，那么倾向于选择内部表，但是如果 Hive 和其他工具要针对相同的数据集进行处理，则外部表更合适。使用外部表访问存储在 HDFS 上的初始数据，然后通过 Hive 转换数据并存到内部表中。使用外部表的场景是针对一个数据集有多个不同的 Schema。通过外部表和内部表的区别和使用选择的对比可以看出，Hive 其实只是对存储在 HDFS 上的数据提供了一种新的抽象，而不是管理存储在 HDFS 上的数据。所以不管创建的是内部表还是外部表，都可以对 Hive 表的数据存储目录中的数据进行增删操作。

2. 分区表和分桶表

Hive 数据表可以根据某些字段进行分区操作，细化数据管理，可以让部分数据查询更快。表和分区也可以进一步被划分为 Buckets，分桶表的原理和 MapReduce 编程中的 HashPartitioner 的原理类似。

分区表和分桶表操作都是细化数据管理，但是分区表是手动添加区分，因为 Hive 是只读模式，所以对添加进分区的数据不做模式校验，分桶表中的数据是按照某些分桶字段进行 Hash 散列形成的多个文件，所以数据的准确性高很多。

5.2 Hive 的安装与使用

5.2.1 Hive 的安装配置

1. Hive 的下载

我们使用下载地址 http://mirrors.hust.edu.cn/apache/，选择合适的 Hive 版本进行下载，进入 stable-2 文件夹可以看到稳定的 2.x 版本是 2.3.8，如图 5-2 所示。

```
/apache/hive/
                File Name              File Size      Date
    ../                                  -              -
    hive-1.2.2/                          -            03-Jul-2020 04:35
    hive-2.3.8/                          -            15-Jan-2021 21:12
    hive-3.1.2/                          -            03-Jul-2020 04:35
    hive-standalone-metastore-3.0.0/     -            03-Jul-2020 04:35
    hive-storage-2.6.1/                  -            03-Jul-2020 04:35
    hive-storage-2.7.2/                  -            03-Jul-2020 04:34
    stable-2/                            -            15-Jan-2021 21:12
```

图 5-2　Hive 的下载

2. Hive 的安装

① 使用 MySQL 作为 Hive 的元数据库，请先行安装 MySQL，在此不再赘述。

② 上传 Hive 安装包，通过远程登录软件或者其他方式将 apache-hive-2.3.8-bin.tar.gz 上传至 hadoop3 虚拟机。

③ 解压安装包：

```
[hadoop@hadoop3 ~]$ tar -zxvf apache-hive-2.3.8-bin.tar.gz -C apps/
```

④ 修改配置文件。配置文件所在目录为 apache-hive-2.3.8-bin/conf：

```
[hadoop@hadoop3 apps]$ cd apache-hive-2.3.8-bin/
[hadoop@hadoop3 apache-hive-2.3.8-bin]$ ls
bin  binary-package-licenses  conf  examples  hcatalog  jdbc  lib  LICENSE  NOTICE  RELEASE_NOTES.txt  scripts
[hadoop@hadoop3 apache-hive-2.3.8-bin]$ cd conf/
[hadoop@hadoop3 conf]$ ls
beeline-log4j2.properties.template      ivysettings.xml
hive-default.xml.template               llap-cli-log4j2.properties.template
hive-env.sh.template                    llap-daemon-log4j2.properties.template
hive-exec-log4j2.properties.template    parquet-logging.properties
hive-log4j2.properties.template
[hadoop@hadoop3 conf]$ pwd
/home/hadoop/apps/apache-hive-2.3.8-bin/conf
[hadoop@hadoop3 conf]$
```

新建 hive-site.xml 并添加以下内容：

```
[hadoop@hadoop3 conf]$ touch hive-site.xml
[hadoop@hadoop3 conf]$ vi hive-site.xml
<configuration>
    <property>
        <name>javax.jdo.option.ConnectionURL</name>
        <value>jdbc:mysql://hadoop1:3306/hivedb?createDatabaseIfNotExist=true</value>
        <description>JDBC connect string</description>
    </property>
    <property>
        <name>javax.jdo.option.ConnectionDriverName</name>
        <value>com.mysql.jdbc.Driver</value>
```

```xml
            <description>Driver for a JDBC</description>
        </property>
        <property>
            <name>javax.jdo.option.ConnectionUserName</name>
            <value>root</value>
            <description>MySQL user</description>
        </property>
        <property>
            <name>javax.jdo.option.ConnectionPassword</name>
            <value>root</value>
            <description>MySQL user password</description>
        </property>
</configuration>
```

以下为可选配置,该配置信息用来指定 Hive 数据仓库的数据在 HDFS 上的存储目录。

```xml
        <property>
            <name>hive.metastore.warehouse.dir</name>
            <value>/hive/warehouse</value>
            <description>hive default warehouse</description>
        </property>
```

⑤ 一定要记得加入 MySQL 驱动包,这里我们选择 mysql-connector-java-5.1.40-bin.jar,该 jar 包放置在 hive 的根路径下的 lib 目录中。

⑥ 安装完成,配置环境变量:

```
[hadoop@hadoop3 lib]$ vi ~/.bashrc
export HIVE_HOME=/home/hadoop/apps/apache-hive-2.3.8-bin
export PATH=$PATH:$HIVE_HOME/bin
```

使修改的配置文件立即生效:

```
[hadoop@hadoop3 lib]$ source ~/.bashrc
```

⑦ 验证 Hive 安装:

```
[hadoop@hadoop3 ~]$ hive --help
Usage ./hive <parameters> --service serviceName <service parameters>
Service List: beeline cleardanglingscratchdir cli hbaseimport hbaseschematool help hiveburninclient hiveserver2 hplsql jar lineage llapdump llap llapstatus metastore metatool orcfiledump rcfilecat schemaTool version
Parameters parsed:
  --auxpath : Auxiliary jars
  --config : Hive configuration directory
  --service : Starts specific service/component. cli is default
Parameters used:
  HADOOP_HOME or HADOOP_PREFIX : Hadoop install directory
  HIVE_OPT : Hive options
For help on a particular service:
```

```
./hive --service serviceName --help
Debug help: ./hive --debug --help
[hadoop@hadoop3 ~]$
```

⑧ 初始化元数据库。

注意：当使用的 Hive 是 2.x 之前的版本时，不做初始化也是可以的，Hive 第一次启动的时候会自动进行初始化，只不过不会生成足够多的元数据库中的表，在使用过程中会慢慢生成，但最后进行初始化。如果使用的是 2.x 版本的 Hive，那么必须手动初始化元数据库。命令如下：

```
[hadoop@hadoop3 ~]$ schematool -dbType mysql -initSchema
SLF4J: Class path contains multiple SLF4J bindings.
SLF4J: Found binding in [jar:file:/home/hadoop/apps/apache-hive-2.3.8-bin/lib/log4j-slf4j-impl-2.6.2.jar!/org/slf4j/impl/StaticLoggerBinder.class]
SLF4J: Found binding in [jar:file:/home/hadoop/apps/hadoop-2.7.5/share/hadoop/common/lib/slf4j-log4j12-1.7.10.jar!/org/slf4j/impl/StaticLoggerBinder.class]
SLF4J: See http://www.slf4j.org/codes.html#multiple_bindings for an explanation.
SLF4J: Actual binding is of type [org.apache.logging.slf4j.Log4jLoggerFactory]
Metastore connection URL:        jdbc:mysql://hadoop1:3306/hivedb?createDatabaseIfNotExist=true
Metastore Connection Driver :    com.mysql.jdbc.Driver
Metastore connection User:       root
Starting metastore schema initialization to 2.3.0
Initialization script hive-schema-2.3.0.mysql.sql
Initialization script completed
schemaTool completed
[hadoop@hadoop3 ~]$
```

⑨ 启动 Hive 客户端，"hive --service cli" 和 "hive" 效果一样：

```
[hadoop@hadoop3 ~]$ hive --service cli
SLF4J: Class path contains multiple SLF4J bindings.
SLF4J: Found binding in [jar:file:/home/hadoop/apps/apache-hive-2.3.8-bin/lib/log4j-slf4j-impl-2.6.2.jar!/org/slf4j/impl/StaticLoggerBinder.class]
SLF4J: Found binding in [jar:file:/home/hadoop/apps/hadoop-2.7.5/share/hadoop/common/lib/slf4j-log4j12-1.7.10.jar!/org/slf4j/impl/StaticLoggerBinder.class]
SLF4J: See http://www.slf4j.org/codes.html#multiple_bindings for an explanation.
SLF4J: Actual binding is of type [org.apache.logging.slf4j.Log4jLoggerFactory]
Logging initialized using configuration in jar:file:/home/Hadoop/apps/apache-hive-2.3.8-bin/lib/hive-common-2.3.3.jar!/hive-log4j2.properties Async: true
Hive-on-MR is deprecated in Hive 2 and may not be available in the future versions. Consider using a different execution engine (i.e. spark, tez) or using Hive 1.X releases.
hive>
```

5.2.2 Hive 的基本使用

我们准备一个案例文件 student.txt,将其存入 Hive 中,student.txt 数据内容如下:

```
95002,刘晨,女,19,IS
95017,王风娟,女,18,IS
95018,王一,女,19,IS
95013,冯伟,男,21,CS
95014,王小丽,女,19,CS
95019,邢小丽,女,19,IS
95020,赵钱,男,21,IS
95003,王敏,女,22,MA
95004,张立,男,19,IS
95012,孙花,女,20,CS
95010,孔小涛,男,19,CS
95005,刘刚,男,18,MA
95006,孙庆,男,23,CS
95007,易思玲,女,19,MA
95008,李娜,女,18,CS
95021,周二,男,17,MA
95022,郑明,男,20,MA
95001,李勇,男,20,CS
95011,包小柏,男,18,MA
95009,梦圆圆,女,18,MA
95015,王君,男,18,MA
```

1. 创建一个数据库 myhive

```
hive> create database myhive;
OK
Time taken: 7.847 seconds
hive>
```

2. 使用新的数据库 myhive

```
hive> use myhive;
OK
Time taken: 0.047 seconds
hive>
```

3. 查看当前正在使用的数据库

```
hive> select current_database();
OK
myhive
Time taken: 0.728 seconds, Fetched: 1 row(s)
hive>
```

4. 在数据库 myhive 中创建一张 student 表

hive> create table student(id int, name string, sex string, age int, department string) row format delimited fields terminated by ",";
OK
Time taken: 0.718 seconds
hive>

5. 向表中加载数据

hive> load data local inpath "/home/hadoop/student.txt" into table student;
Loading data to table myhive.student
OK
Time taken: 1.854 seconds
hive>

6. 查询数据

hive> select * from student;
OK
95002	刘晨	女	19	IS
95017	王风娟	女	18	IS
95018	王一	女	19	IS
95013	冯伟	男	21	CS
95014	王小丽	女	19	CS
95019	邢小丽	女	19	IS
95020	赵钱	男	21	IS
95003	王敏	女	22	MA
95004	张立	男	19	IS
95012	孙花	女	20	CS
95010	孔小涛	男	19	CS
95005	刘刚	男	18	MA
95006	孙庆	男	23	CS
95007	易思玲	女	19	MA
95008	李娜	女	18	CS
95021	周二	男	17	MA
95022	郑明	男	20	MA
95001	李勇	男	20	CS
95011	包小柏	男	18	MA
95009	梦圆圆	女	18	MA
95015	王君	男	18	MA

Time taken: 2.455 seconds, Fetched: 21 row(s)
hive>

7. 查看表结构

hive> desc student;
OK
id   int

name	string
sex	string
age	int
department	string

Time taken: 0.102 seconds, Fetched: 5 row(s)
hive>

hive> desc extended student;

OK
id	int
name	string
sex	string
age	int
department	string

Detailed Table Information   Table(tableName:student, dbName:myhive, owner:Hadoop, createTime:1522750487, lastAccessTime:0, retention:0, sd:StorageDescriptor(cols:[FieldSchema(name: id, type:int, comment:null), FieldSchema(name:name, type:string, comment:null), FieldSchema(name: sex, type:string, comment:null), FieldSchema(name:age, type:int, comment:null), FieldSchema(name: department, type:string, comment:null)], location:hdfs://hadoop1:9000/user/hive/warehouse/myhive. db/student, inputFormat:org.apache.hadoop.mapred.TextInputFormat, outputFormat:org.apache.hadoop. hive.ql.io.HiveIgnoreKeyTextOutputFormat, compressed:false, numBuckets:-1, serdeInfo:SerDeInfo (name:null, serializationLib:org.apache.Hadoop.hive.serde2.lazy.LazySimpleSerDe, parameters: {serialization.format=,, field.delim=,}), bucketCols:[], sortCols:[], parameters:{}, skewedInfo: SkewedInfo(skewedColNames:[], skewedColValues:[], skewedColValueLocationMaps:{}), storedAsSubDirectories:false), partitionKeys:[], parameters:{transient_lastDdlTime=1522750695, totalSize=523, numRows=0, rawDataSize=0, numFiles=1}, viewOriginalText:null, viewExpandedText: null, tableType:MANAGED_TABLE, rewriteEnabled:false)

Time taken: 0.127 seconds, Fetched: 7 row(s)
hive>

hive> desc formatted student;

OK
# col_name	data_type	comment
id	int	
name	string	
sex	string	
age	int	
department	string	
# Detailed Table Information		
Database:	myhive	
Owner:	hadoop	
CreateTime:	Tue Apr 08 18:14:47 CST 2020	
LastAccessTime:	UNKNOWN	
Retention:	0	
Location:	hdfs://hadoop1:9000/user/hive/warehouse/myhive.db/student	
Table Type:	MANAGED_TABLE	
Table Parameters:		
numFiles	1	

```
    numRows                    0
    rawDataSize                0
    totalSize                  523
    transient_lastDdlTime      1522750695
# Storage Information
SerDe Library:                 org.apache.hadoop.hive.serde2.lazy.LazySimpleSerDe
InputFormat:                   org.apache.hadoop.mapred.TextInputFormat
OutputFormat:                  org.apache.hadoop.hive.ql.io.HiveIgnoreKeyTextOutputFormat
Compressed:                    No
Num Buckets:                   -1
Bucket Columns:                []
Sort Columns:                  []
Storage Desc Params:
    field.delim                ,
    serialization.format       ,
Time taken: 0.13 seconds, Fetched: 34 row(s)
hive>
```

5.2.3 Hive 的连接方式

1. CLI 连接方式

进入 bin 目录下,直接输入以下命令:

```
[hadoop@hadoop3 ~]$ hive
SLF4J: Class path contains multiple SLF4J bindings.
SLF4J: Found binding in [jar:file:/home/hadoop/apps/apache-hive-2.3.8-bin/lib/log4j-slf4j-impl-2.6.2.jar!/org/slf4j/impl/StaticLoggerBinder.class]
SLF4J: Found binding in [jar:file:/home/hadoop/apps/hadoop-2.7.5/share/hadoop/common/lib/slf4j-log4j12-1.7.10.jar!/org/slf4j/impl/StaticLoggerBinder.class]
SLF4J: See http://www.slf4j.org/codes.html#multiple_bindings for an explanation.
SLF4J: Actual binding is of type [org.apache.logging.slf4j.Log4jLoggerFactory]
Logging initialized using configuration in jar:file:/home/hadoop/apps/apache-hive-2.3.8-bin/lib/hive-common-2.3.3.jar!/hive-log4j2.properties Async: true
Hive-on-MR is deprecated in Hive 2 and may not be available in the future versions. Consider using a different execution engine (i.e. spark, tez) or using Hive 1.X releases.
hive> show databases;
OK
default
myhive
Time taken: 6.569 seconds, Fetched: 2 row(s)
hive>
```

启动成功后,我们接下来便可以做 hive 相关操作。

2. HiveServer2/Beeline

在 Hive 2.3.8 版本中,需要对 Hadoop 集群做如下改变,否则无法使用。

① 修改 Hadoop 集群的 hdfs-site.xml 配置文件,加入一条配置信息,表示启用 webhdfs

功能。

```
<property>
    <name>dfs.webhdfs.enabled</name>
    <value>true</value>
</property>
```

② 修改 Hadoop 集群的 core-site.xml 配置文件，加入两条配置信息，用来表示设置 hadoop 代理用户。

```
<property>
    <name>hadoop.proxyuser.hadoop.hosts</name>
    <value>*</value>
</property>
<property>
    <name>hadoop.proxyuser.hadoop.groups</name>
    <value>*</value>
</property>
```

我们来对配置进行解析：hadoop.proxyuser.hadoop.hosts 配置成 * 的意义是任意节点使用 Hadoop 集群的代理用户 hadoop 都能访问 HDFS 集群，hadoop.proxyuser.hadoop.groups 表示代理用户的组所属。

以上操作完成之后，建议重启 HDFS 集群，然后继续做如下两步。

第一步：启动 HiveServer2 服务。

```
[hadoop@hadoop3 ~]$ hiveserver2
2020-04-14 11:34:49: Starting HiveServer2
SLF4J: Class path contains multiple SLF4J bindings.
SLF4J: Found binding in [jar:file:/home/hadoop/apps/apache-hive-2.3.8-bin/lib/log4j-slf4j-impl-2.6.2.jar!/org/slf4j/impl/StaticLoggerBinder.class]
SLF4J: Found binding in [jar:file:/home/hadoop/apps/hadoop-2.7.5/share/hadoop/common/lib/slf4j-log4j12-1.7.10.jar!/org/slf4j/impl/StaticLoggerBinder.class]
SLF4J: See http://www.slf4j.org/codes.html#multiple_bindings for an explanation.
SLF4J: Actual binding is of type [org.apache.logging.slf4j.Log4jLoggerFactory]
```

此时启动会多一个进程 Runjar，表示 HiveServer2 进程启动。

如果我们需要启动为后台服务，可以尝试使用如下几种方式来实现：

- nohup hiveserver2 1>/home/hadoop/hiveserver.log 2>/home/hadoop/hiveserver.err &；
- nohup hiveserver2 1>/dev/null 2>/dev/null &；
- nohup hiveserver2 >/dev/null 2>&1 &。

以上 3 个命令是等价的，第一个表示记录日志，第二个和第三个表示不记录日志。如果没有配置日志的输出路径，日志会生成在当前工作目录中，默认的日志名称为 nohup.xxx。

```
[hadoop@hadoop3 ~]$ nohup hiveserver2 1>/home/hadoop/log/hivelog/hiveserver.log 2>/home/hadoop/log/hivelog/hiveserver.err &
```

第二步：启动 Beeline 客户端去连接，执行如下命令，其中参数 -u 表示指定元数据库的链接信息，-n 表示指定用户名和密码。

```
[hadoop@hadoop3 ~]$ beeline -u jdbc:hive2//hadoop3:10000 -n hadoop
SLF4J: Class path contains multiple SLF4J bindings.
SLF4J: Found binding in [jar:file:/home/hadoop/apps/apache-hive-2.3.8-bin/lib/log4j-slf4j-impl-2.6.2.jar!/org/slf4j/impl/StaticLoggerBinder.class]
SLF4J: Found binding in [jar:file:/home/hadoop/apps/hadoop-2.7.5/share/hadoop/common/lib/slf4j-log4j12-1.7.10.jar!/org/slf4j/impl/StaticLoggerBinder.class]
SLF4J: See http://www.slf4j.org/codes.html#multiple_bindings for an explanation.
SLF4J: Actual binding is of type [org.apache.logging.slf4j.Log4jLoggerFactory]
scan complete in 1ms
scan complete in 2374ms
No known driver to handle "jdbc:hive2//hadoop3:10000"
Beeline version 2.3.3 by Apache Hive
beeline>
```

还有一种方式也可以去连接,先执行"beeline",然后在提示符中输入"!connect jdbc:hive2://hadoop02:10000",按"Enter"键,然后输入用户名,这个用户名就是安装 Hadoop 集群的用户名,接下来可以进行 hive 操作。

```
[hadoop@hadoop3 ~]$ beeline
SLF4J: Class path contains multiple SLF4J bindings.
SLF4J: Found binding in [jar:file:/home/hadoop/apps/apache-hive-2.3.8-bin/lib/log4j-slf4j-impl-2.6.2.jar!/org/slf4j/impl/StaticLoggerBinder.class]
SLF4J: Found binding in [jar:file:/home/hadoop/apps/hadoop-2.7.5/share/hadoop/common/lib/slf4j-log4j12-1.7.10.jar!/org/slf4j/impl/StaticLoggerBinder.class]
SLF4J: See http://www.slf4j.org/codes.html#multiple_bindings for an explanation.
SLF4J: Actual binding is of type [org.apache.logging.slf4j.Log4jLoggerFactory]
Beeline version 2.3.8 by Apache Hive
beeline> ! connect jdbc:hive2://hadoop3:10000
Connecting to jdbc:hive2://hadoop3:10000
Enter username for jdbc:hive2://hadoop3:10000: hadoop
Enter password for jdbc:hive2://hadoop3:10000: ******
Connected to: Apache Hive (version 2.3.8)
Driver: Hive JDBC (version 2.3.8)
Transaction isolation: TRANSACTION_REPEATABLE_READ
0: jdbc:hive2://hadoop3:10000>
```

5.3 Hive 数据结构

5.3.1 Hive 数据类型

1. 基本数据类型

Hive 支持关系型数据中的大多数基本数据类型,可以参考表 5-1。

表 5-1 Hive 支持的基本数据类型

类型	描述	示例
boolean	true/false	TRUE
tinyint	1字节有符号整数，-128~127	1Y
smallint	2字节有符号整数，-32768~32767	1S
int	4字节有符号整数	1
bigint	8字节有符号整数	1L
float	4字节单精度浮点数	1.0
double	8字节双精度浮点数	1.0
deicimal	任意精度的有符号小数	1.0
string	变长字符串	"a",'b'
varchar	变长字符串	"a",'b'
char	固定长度字符串	"a",'b'
binary	字节数组	无法表示
timestamp	时间戳,纳秒精度	122327493795
date	日期	'2018-04-07'

和其他的 SQL 一样,这些都是保留字。需要注意的是,所有的这些数据类型都是对 Java 中接口的实现,因此这些类型的具体行为细节和 Java 中对应的类型是完全一致的。例如,string 类型实现的是 Java 中的 string,float 实现的是 Java 中的 float,等等。

2. 复杂数据类型

Hive 支持的复杂数据类型如表 5-2 所示。

表 5-2 Hive 支持的复杂数据类型

类型	描述	示例
array	有序的同类型的集合	array(1,2)
map	key-value，key 必须为原始类型,value 可以为任意类型	map('a',1,'b',2)
struct	字段集合,类型可以不同	struct('1',1,1.0), named_struct('col1','1','col2',1,'clo3',1.0)

5.3.2 Hive 数据存储格式

Hive 会为每个创建的数据库在 HDFS 上创建一个目录,该数据库的表会以子目录形式存储,表中的数据会以表目录下的文件形式存储。Default 数据库没有自己的目录,Default 数据库中的表默认存放在 /user/hive/warehouse 目录下。

1. TextFile

TextFile 为默认格式,存储方式为行存储。数据不做压缩,磁盘开销大,数据解析开销大。

2. SequenceFile

SequenceFile 是 Hadoop API 提供的一种二进制文件支持,其具有使用方便、可分割、可压缩的特点。SequenceFile 支持 3 种压缩选择：NONE、RECORD、BLOCK。RECORD 压缩率低，一般建议使用 BLOCK 压缩。

3. RCFile

一种行列存储相结合的存储方式。

4. ORCFile

数据按照行分块，每个块按照列进行存储，其中每个块都存储有一个索引。Hive 给出的新格式，属于 RCFile 的升级版，性能有大幅度提升，而且数据可以压缩存储，快速列存取。

5. Parquet

Parquet 也是一种行式存储，具有很好的压缩性能，同时可以减少大量的表扫描和反序列化的时间。

5.3.3 数据格式

当数据存储在文本文件中，必须按照一定格式区分行和列，并且在 Hive 中指明分隔符。Hive 默认使用了几个平时很少出现的字符，这些字符一般不会作为内容出现在记录中。Hive 默认的行和列分隔符如表 5-3 所示。

表 5-3 Hive 默认的行和列分隔符

分隔符	描述
\n	对于文本文件来说，每行是一条记录，所以用\n 来分割记录
^A (Ctrl+A)	分割字段，也可以用\001 表示
^B(Ctrl+B)	用于分割 Array 或 Struct 中的元素，或者用于分割 Map 中的键和值，也可以用\002 表示
^C	用于 Map 中键和值的分割，也可以用\003 表示

5.4 Hive 数据操作

5.4.1 管理库

1. 创建数据仓库

参考 Hive 官网，创建数据仓库的语法结构如下：

```
CREATE (DATABASE|SCHEMA) [IF NOT EXISTS] database_name
    [COMMENT database_comment]  //关于数据块的描述
    [LOCATION hdfs_path]         //指定数据库在 HDFS 上的存储位置
    [WITH DBPROPERTIES (property_name = property_value, ...)]; //指定数据块属性
```

默认数据地址：/user/hive/warehouse/db_name.db/table_name/partition_name/…。

（1）创建普通的数据库

```
0: jdbc:hive2://hadoop3:10000> create database t1;
```

```
No rows affected (0.308 seconds)
0：jdbc：hive2：//hadoop3：10000 > show databases；
+ ------------------ +
| database_name |
+ ------------------ +
| default |
| myhive |
| t1 |
+ ------------------ +
3 rows selected (0.393 seconds)
0：jdbc：hive2：//hadoop3：10000 >
```

（2）创建库的时候检查存在与否

```
0：jdbc：hive2：//hadoop3：10000 > create database if not exists t1；
No rows affected (0.176 seconds)
0：jdbc：hive2：//hadoop3：10000 >
```

（3）创建库的时候带注释

```
0：jdbc：hive2：//hadoop3：10000 > create database if not exists t2 comment 'learning hive'；
No rows affected (0.217 seconds)
0：jdbc：hive2：//hadoop3：10000 >
```

（4）创建带属性的库

```
0：jdbc：hive2：//hadoop3：10000 > create database if not exists t3 with dbproperties('creator'='hadoop','date'='2020-04-08')；
No rows affected (0.255 seconds)
0：jdbc：hive2：//hadoop3：10000 >
```

2．查看数据仓库

（1）查看有哪些数据库

```
0：jdbc：hive2：//hadoop3：10000 > show databases；
+ ------------------ +
| database_name |
+ ------------------ +
| default |
| myhive |
| t1 |
| t2 |
| t3 |
+ ------------------ +
5 rows selected (0.164 seconds)
0：jdbc：hive2：//hadoop3：10000 >
```

（2）显示数据库的详细属性信息

语法结构如下：

```
desc database [extended] dbname；
```

参考示例如下：

```
0: jdbc:hive2://hadoop3:10000> desc database extended t3;
+----------+----------+--------------------------------------------------+------------+------------+
| db_name  | comment  |                     location                     | owner_name | owner_type |
|          parameters         |
+----------+----------+--------------------------------------------------+------------+------------+
| t3       |          | hdfs://hadoop1:9000/user/hive/warehouse/t3.db    | hadoop     | USER       |
|     {date=2020-04-08, creator=hadoop}     |
+----------+----------+--------------------------------------------------+------------+------------+
1 row selected (0.11 seconds)
0: jdbc:hive2://hadoop3:10000>
```

（3）查看正在使用哪个库

```
0: jdbc:hive2://hadoop3:10000> select current_database();
+----------+
|   _c0    |
+----------+
| default  |
+----------+
1 row selected (1.36 seconds)
0: jdbc:hive2://hadoop3:10000>
```

（4）查看创建库的详细语句

```
0: jdbc:hive2://hadoop3:10000> show create database t3;
+----------------------------------------------------+
|                  createdb_stmt                     |
+----------------------------------------------------+
| CREATE DATABASE t3                                 |
| LOCATION                                           |
|   'hdfs://hadoop1:9000/user/hive/warehouse/t3.db'  |
| WITH DBPROPERTIES (                                |
|   'creator'='hadoop',                              |
|   'date'='2020-04-08')                             |
+----------------------------------------------------+
6 rows selected (0.155 seconds)
0: jdbc:hive2://hadoop3:10000>
```

3. 删除数据仓库

删除数据仓库的操作可以参考如下命令：

- drop database dbname;
- drop database if exists dbname。

默认情况下，Hive 不允许删除包含表的数据库，有以下两种解决办法。

- 手动删除库下所有表,然后删除库。
- 使用 cascade 关键字:drop database if exists dbname cascade。

默认情况下就是"restrict drop database if exists myhive"操作,等同于"drop database if exists myhive restrict"操作。

(1) 删除不含表的数据库

```
0:jdbc:hive2://hadoop3:10000> show tables in t1;
+----------+
| tab_name |
+----------+
+----------+
No rows selected (0.147 seconds)
0:jdbc:hive2://hadoop3:10000> drop database t1;
No rows affected (0.178 seconds)
0:jdbc:hive2://hadoop3:10000> show databases;
+----------------+
| database_name  |
+----------------+
| default        |
| myhive         |
| t2             |
| t3             |
+----------------+
4 rows selected (0.124 seconds)
0:jdbc:hive2://hadoop3:10000>
```

(2) 删除含有表的数据库

```
0:jdbc:hive2://hadoop3:10000> drop database if exists t3 cascade;
No rows affected (1.56 seconds)
0:jdbc:hive2://hadoop3:10000>
```

4. 切换数据仓库

语法结构如下:

```
use database_name
```

参考示例如下:

```
0:jdbc:hive2://hadoop3:10000> use t2;
No rows affected (0.109 seconds)
0:jdbc:hive2://hadoop3:10000>
```

5.4.2 表操作

1. 创建表

从 Hive 官网中查询到创建表的语法结构如下:

```
CREATE [EXTERNAL] TABLE [IF NOT EXISTS] table_name
    [(col_name data_type [COMMENT col_comment],...)]
    [COMMENT table_comment]
    [PARTITIONED BY (col_name data_type [COMMENT col_comment],...)]
    [CLUSTERED BY (col_name, col_name,...)
        [SORTED BY (col_name [ASC|DESC],...)] INTO num_buckets BUCKETS]
    [ROW FORMAT row_format]
    [STORED AS file_format]
    [LOCATION hdfs_path]
```

语法结构中相关参数解释如下。

CREATE TABLE：创建一个指定名字的表。如果相同名字的表已经存在，则抛出异常；用户可以用 IF NOT EXISTS 选项来忽略这个异常。

EXTERNAL：该关键字可以让用户创建一个外部表，在建表的同时指定一个指向实际数据的路径（LOCATION）。

LIKE：允许用户复制现有的表结构，但是不复制数据。

COMMENT：可以为表与字段增加描述。

PARTITIONED BY：指定分区。

```
ROW FORMAT DELIMITED [FIELDS TERMINATED BY char] [COLLECTION ITEMS TERMINATED BY char] MAP KEYS
TERMINATED BY char] [LINES TERMINATED BY char] | SERDE serde_name [WITH SERDEPROPERTIES (property_name =
property_value, property_name = property_value,...)]
```

用户在建表的时候可以自定义 SerDe 或者使用自带的 SerDe，如果没有指定 ROW FORMAT 或者 ROW FORMAT DELIMITED，将会使用自带的 SerDe。在建表的时候，用户还需要为表指定列，用户在指定表的列的同时也会指定自定义的 SerDe，Hive 通过 SerDe 确定表的具体列的数据。

```
STORED AS
    SEQUENCEFILE  //序列化文件
    | TEXTFILE    //普通的文本文件格式
    | RCFILE      //行列存储相结合的文件
    | INPUTFORMAT input_format_classname OUTPUTFORMAT output_format_classname //自定义文件格式
```

如果文件数据是纯文本，可以使用 STORED AS TEXTFILE。如果数据需要压缩，则使用 STORED AS SEQUENCEFILE。

LOCATION 指定表在 HDFS 上的存储路径，如果一份数据已经存储到 HDFS 上，并且要被多个用户或者客户端使用，最好创建外部表，反之最好创建内部表。如果不指定，则按照默认的规则存储在默认的仓库路径中。

示例：使用 t2 数据库进行操作。

（1）创建默认的内部表

```
0: jdbc:hive2://hadoop3:10000 > create table student(id int, name string, sex string, age int,
department string) row format delimited fields terminated by ",";
No rows affected (0.222 seconds)
0: jdbc:hive2://hadoop3:10000 > desc student;
```

```
+-------------+-----------+---------+
|  col_name   | data_type | comment |
+-------------+-----------+---------+
| id          | int       |         |
| name        | string    |         |
| sex         | string    |         |
| age         | int       |         |
| department  | string    |         |
+-------------+-----------+---------+
5 rows selected (0.168 seconds)
0: jdbc:hive2://hadoop3:10000 >
```

(2)创建外部表

```
0: jdbc:hive2://hadoop3:10000 > create external table student_ext(id int, name string, sex string, age int, department string) row format delimited fields terminated by "," location "/hive/student";
No rows affected (0.248 seconds)
0: jdbc:hive2://hadoop3:10000 >
```

(3)创建分区表

```
0: jdbc:hive2://hadoop3:10000 > create external table student_ptn(id int, name string, sex string, age int, department string) partitioned by (city string) row format delimited fields terminated by "," location "/hive/student_ptn";
No rows affected (0.24 seconds)
0: jdbc:hive2://hadoop3:10000 >
```

添加分区:

```
0: jdbc:hive2://hadoop3:10000 > alter table student_ptn add partition(city = "beijing");
No rows affected (0.269 seconds)
0: jdbc:hive2://hadoop3:10000 > alter table student_ptn add partition(city = "shenzhen");
No rows affected (0.236 seconds)
0: jdbc:hive2://hadoop3:10000 >
```

如果某张表是分区表,那么每个分区的定义其实就表现为这张表的数据存储目录下的一个子目录。如果是分区表,那么数据文件一定要存储在某个分区中,而不能直接存储在表中。

(4)创建分桶表

```
0: jdbc:hive2://hadoop3:10000 > create external table student_bck(id int, name string, sex string, age int, department string) clustered by (id) sorted by (id asc, name desc) into 4 buckets row format delimited fields terminated by "," location "/hive/student_bck";
No rows affected (0.216 seconds)
0: jdbc:hive2://hadoop3:10000 >
```

(5)使用 CTAS 创建表

使用 CTAS 创建表就是由一个查询 SQL 的结果来创建一张表进行存储。现向 student 表中导入数据,使用/home/hadoop/student.txt 文件作为数据源。

```
0: jdbc:hive2://hadoop3:10000 > load data local inpath "/home/hadoop/student.txt" into table student;
No rows affected (0.715 seconds)
0: jdbc:hive2://hadoop3:10000 > select * from student;
```

student.id	student.name	student.sex	student.age	student.department
95002	刘晨	女	19	IS
95017	王风娟	女	18	IS
95018	王一	女	19	IS
95013	冯伟	男	21	CS
95014	王小丽	女	19	CS
95019	邢小丽	女	19	IS
95020	赵钱	男	21	IS
95003	王敏	女	22	MA
95004	张立	男	19	IS
95012	孙花	女	20	CS
95010	孔小涛	男	19	CS
95005	刘刚	男	18	MA
95006	孙庆	男	23	CS
95007	易思玲	女	19	MA
95008	李娜	女	18	CS
95021	周二	男	17	MA
95022	郑明	男	20	MA
95001	李勇	男	20	CS
95011	包小柏	男	18	MA
95009	梦圆圆	女	18	MA
95015	王君	男	18	MA

```
21 rows selected (0.342 seconds)
0: jdbc:hive2://hadoop3:10000 >
```

使用 CTAS 创建表：

```
0: jdbc:hive2://hadoop3:10000 > create table student_ctas as select * from student where id < 95012;
WARNING: Hive-on-MR is deprecated in Hive 2 and may not be available in the future versions. Consider using a different execution engine (i.e. spark, tez) or using Hive 1.X releases.
No rows affected (34.514 seconds)
0: jdbc:hive2://hadoop3:10000 > select * from student_ctas;
```

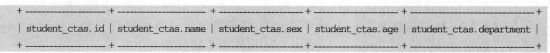

student_ctas.id	student_ctas.name	student_ctas.sex	student_ctas.age	student_ctas.department

```
| 95002        | 刘晨         | 女           | 19           | IS           |
| 95003        | 王敏         | 女           | 22           | MA           |
| 95004        | 张立         | 男           | 19           | IS           |
| 95010        | 孔小涛       | 男           | 19           | CS           |
| 95005        | 刘刚         | 男           | 18           | MA           |
| 95006        | 孙庆         | 男           | 23           | CS           |
| 95007        | 易思玲       | 女           | 19           | MA           |
| 95008        | 李娜         | 女           | 18           | CS           |
| 95001        | 李勇         | 男           | 20           | CS           |
| 95011        | 包小柏       | 男           | 18           | MA           |
| 95009        | 梦圆圆       | 女           | 18           | MA           |
+--------------+--------------+--------------+--------------+--------------+
11 rows selected (0.445 seconds)
0: jdbc:hive2://hadoop3:10000 >
```

(6) 复制表结构

```
0: jdbc:hive2://hadoop3:10000 > create table student_copy like student;
No rows affected (0.217 seconds)
0: jdbc:hive2://hadoop3:10000 >
```

注意：如果在 table 的前面没有加 external 关键字，那么复制出来的新表无论如何都是内部表；如果在 table 的前面加了 external 关键字，那么复制出来的新表无论如何都是外部表。

2. 查看表

(1) 查看表列表

查看当前使用的数据库中有哪些表，可以使用以下命令：

```
0: jdbc:hive2://hadoop3:10000 > show tables;
+------------------+
|    tab_name      |
+------------------+
| student          |
| student_bck      |
| student_copy     |
| student_ctas     |
| student_ext      |
| student_ptn      |
+------------------+
6 rows selected (0.163 seconds)
0: jdbc:hive2://hadoop3:10000 >
```

查看其他数据仓库中有哪些表，可以使用以下命令：

```
0: jdbc:hive2://hadoop3:10000 > show tables in myhive;
```

```
+------------+
|  tab_name  |
+------------+
|  student   |
+------------+
1 row selected (0.144 seconds)
0: jdbc:hive2://hadoop3:10000>
```

查看数据库中以 student 开头的表,我们可以使用模糊查询 like:

```
0: jdbc:hive2://hadoop3:10000> show tables like 'student_c*';
+---------------+
|   tab_name    |
+---------------+
| student_copy  |
| student_ctas  |
+---------------+
2 rows selected (0.13 seconds)
0: jdbc:hive2://hadoop3:10000>
```

(2) 查看表的详细信息

查看表的信息,进行表的描述性统计,可以使用如下命令:

```
0: jdbc:hive2://hadoop3:10000> desc student;
+-------------+-------------+----------+
|  col_name   |  data_type  | comment  |
+-------------+-------------+----------+
| id          | int         |          |
| name        | string      |          |
| sex         | string      |          |
| age         | int         |          |
| department  | string      |          |
+-------------+-------------+----------+
5 rows selected (0.149 seconds)
0: jdbc:hive2://hadoop3:10000>
```

查看表的详细信息,可以使用"desc extended 表名",但是显示结果排版比较松垮,界面不友好,示例如下:

```
0: jdbc:hive2://hadoop3:10000> desc extended student;
OK
id                      int
name                    string
sex                     string
age                     int
department              string
Detailed Table Information Table(tableName:student, dbName:t2, owner:hadoop, createTime:1547635948, lastAccessTime:0, retention:0, sd:StorageDescriptor(cols:[FieldSchema(name:id, type:
```

int, comment:null), FieldSchema(name:name, type:string, comment:null), FieldSchema(name:sex, type: string, comment:null), FieldSchema(name:age, type:int, comment:null), FieldSchema(name:department, type:string, comment: null)], location:hdfs://localhost:9000/user/hive/warehouse/t2.db/student, inputFormat:org.apache.hadoop.mapred.TextInputFormat, outputFormat:org.apache.hadoop.hive.ql.io. HiveIgnoreKeyTextOutputFormat, compressed:false, numBuckets:-1, serdeInfo:SerDeInfo(name:null, serializationLib:org.apache.hadoop.hive.serde2.lazy.LazySimpleSerDe, parameters:{serialization. format =,, field.delim =,}), bucketCols:[], sortCols:[], parameters:{}, skewedInfo:SkewedInfo (skewedColNames:[], skewedColValues:[], skewedColValueLocationMaps:{}), storedAsSubDirectories: false), partitionKeys:[], parameters:{transient_lastDdlTime = 1547691275, totalSize = 531, numRows = 0, rawDataSize = 0, numFiles = 2}, viewOriginalText:null, viewExpandedText:null, tableType:MANAGED_ TABLE)

Time taken: 3.138 seconds, Fetched: 7 row(s)

查看表的详细信息，还可以使用"desc formatted 表名"，显示结果比较友好，示例如下：

0: jdbc:hive2://hadoop3:10000 > desc formatted student;
OK

# col_name	data_type	comment
id	int	
name	string	
sex	string	
age	int	
department	string	
# Detailed Table Information		
Database:	t2	
Owner:	hadoop	
CreateTime:	Wed Jan 16 18:52:28 CST 2019	
LastAccessTime:	UNKNOWN	
Retention:	0	
Location:	hdfs://localhost:9000/user/hive/warehouse/t2.db/student	
Table Type:	MANAGED_TABLE	
Table Parameters:		
numFiles	2	
numRows	0	
rawDataSize	0	
totalSize	531	
transient_lastDdlTime	1547691275	
# Storage Information		
SerDe Library:	org.apache.hadoop.hive.serde2.lazy.LazySimpleSerDe	
InputFormat:	org.apache.hadoop.mapred.TextInputFormat	
OutputFormat:	org.apache.hadoop.hive.ql.io.HiveIgnoreKeyTextOutputFormat	
Compressed:	No	
Num Buckets:	-1	
Bucket Columns:	[]	
Sort Columns:	[]	
Storage Desc Params:		
field.delim	,	

```
    serialization.format    ,
Time taken: 0.615 seconds, Fetched: 34 row(s)
```

查看分区信息,使用"show partitions 表名",示例如下:

```
0: jdbc:hive2://hadoop3:10000 > show partitions student_ptn;
OK
city = beijing
city = shenzhen
Time taken: 1.437 seconds, Fetched: 2 row(s)
```

(3) 查看表的详细建表语句

```
0: jdbc:hive2://hadoop3:10000 > show create table student_ptn;
OK
CREATE EXTERNAL TABLE student_ptn(
  id int,
  name string,
  sex string,
  age int,
  department string)
PARTITIONED BY (
  city string)
ROW FORMAT SERDE
  'org.apache.hadoop.hive.serde2.lazy.LazySimpleSerDe'
WITH SERDEPROPERTIES (
  'field.delim'=',',
  'serialization.format'=',')
STORED AS INPUTFORMAT
  'org.apache.hadoop.mapred.TextInputFormat'
OUTPUTFORMAT
  'org.apache.hadoop.hive.ql.io.HiveIgnoreKeyTextOutputFormat'
LOCATION
  'hdfs://hadoop1:9000/hive/student_ptn'
TBLPROPERTIES (
  'transient_lastDdlTime'='1547636036')
Time taken: 2.564 seconds, Fetched: 21 row(s)
```

3. 修改表

(1) 修改表名

```
0: jdbc:hive2://hadoop3:10000 > alter table student rename to new_student;
OK
Time taken: 1.763 seconds
0: jdbc:hive2://hadoop3:10000 > show tables;
OK
new_student
student_bck
```

student_copy
student_ext
student_ptn
Time taken: 0.355 seconds, Fetched: 5 row(s)

(2) 修改字段定义

① 增加一个字段

0: jdbc:hive2://hadoop3:10000 > alter table new_student add columns (score int);

OK
Time taken: 1.844 seconds

0: jdbc:hive2://hadoop3:10000 > desc new_student;

OK
id	int
name	string
sex	string
age	int
department	string
score	int

Time taken: 0.374 seconds, Fetched: 6 row(s)

② 修改一个字段的定义

0: jdbc:hive2://hadoop3:10000 > alter table new_student change name new_name string;

OK
Time taken: 4.538 seconds

0: jdbc:hive2://hadoop3:10000 > desc new_student;

OK
id	int
new_name	string
sex	string
age	int
department	string
score	int

Time taken: 0.97 seconds, Fetched: 6 row(s)

③ 删除一个字段

Hive 不支持删除表字段，Hive 中不支持"alter table table_name drop columns"这种语法，支持替换字段。

④ 替换所有字段

0: jdbc:hive2://hadoop3:10000 > alter table new_student replace columns (id int, name string, address string);

OK
Time taken: 1.492 seconds

0: jdbc:hive2://hadoop3:10000 > desc new_student;

```
OK
id                      int
name                    string
address                 string
Time taken: 0.651 seconds, Fetched: 3 row(s)
```

(3) 修改分区信息

① 添加分区:添加一个静态分区

```
0: jdbc:hive2://hadoop3:10000 > alter table student_ptn add partition(city = "shanghai");
OK
Time taken: 1.892 seconds
```

```
0: jdbc:hive2://hadoop3:10000 > alter table student_ptn add partition(city = "guangzhou") partition(city = "chongqing") partition(city = "nanjing");
OK
Time taken: 1.599 seconds
```

先向 student_ptn 表中插入数据,数据格式如下:

```
0: jdbc:hive2://hadoop3:10000 > load data local inpath "/home/hadoop/student.txt" into table student_ptn partition(city = "beijing");
Loading data to table t2.student_ptn partition (city = beijing)
OK
Time taken: 8.856 seconds
```

```
hive > select * from student_ptn;
OK
95002    刘晨      女    19    IS    beijing
95017    王风娟    女    18    IS    beijing
95018    王一      女    19    IS    beijing
95013    冯伟      男    21    CS    beijing
95014    王小丽    女    19    CS    beijing
95019    邢小丽    女    19    IS    beijing
95020    赵钱      男    21    IS    beijing
95003    王敏      女    22    MA    beijing
95004    张立      男    19    IS    beijing
95012    孙花      女    20    CS    beijing
95010    孔小涛    男    19    CS    beijing
95005    刘刚      男    18    MA    beijing
95006    孙庆      男    23    CS    beijing
95007    易思玲    女    19    MA    beijing
95008    李娜      女    18    CS    beijing
95021    周二      男    17    MA    beijing
95022    郑明      男    20    MA    beijing
95001    李勇      男    20    CS    beijing
95011    包小柏    男    18    MA    beijing
95009    梦圆圆    女    18    MA    beijing
95015    王君      男    18    MA    beijing
Time taken: 0.916 seconds, Fetched: 21 row(s)
```

现在我们把这张表的内容直接插入另一张表 student_ptn_age 中,并实现 sex 为动态分区(不指定到底是哪种性别,让系统自己分配决定)。首先创建 student_ptn_age 并指定分区为 age:

```
0: jdbc:hive2://hadoop3:10000 > create table student_ptn_age(id int,name string,sex string,department string) partitioned by (age int);
OK
Time taken: 1.025 seconds
```

然后从 student_ptn 表中查询数据并插入 student_ptn_age 表中:

```
0: jdbc:hive2://hadoop3:10000 > insert overwrite table student_ptn_age partition(age) select id,name,sex,department,age from student_ptn;
WARNING: Hive-on-MR is deprecated in Hive 2 and may not be available in the future versions. Consider using a different execution engine (i.e. spark, tez) or using Hive 1.X releases.
No rows affected (27.905 seconds)
0: jdbc:hive2://hadoop3:10000 >
```

② 修改分区

修改分区一般是指修改分区的数据存储目录。在添加分区的时候,可直接指定当前分区的数据存储目录:

```
0: jdbc:hive2://hadoop3:10000 > alter table student_ptn add if not exists partition(city='beijing') location '/student_ptn_beijing' partition(city='HK') location '/student_HK';
No rows affected (0.306 seconds)
0: jdbc:hive2://hadoop3:10000 >
```

修改已经指定好的分区的数据存储目录:

```
0: jdbc:hive2://hadoop3:10000 > alter table student_ptn partition (city='beijing') set location '/student_ptn_beijing';
OK
Time taken: 1.913 seconds
```

此时原先的分区文件夹仍存在,但是在向分区中添加数据时,只会添加到新的分区目录。

③ 删除分区

```
0: jdbc:hive2://hadoop3:10000 > alter table student_ptn drop partition (city='beijing');
Dropped the partition city=beijing
OK
Time taken: 3.197 seconds
```

删除后再次查看 student_ptn 表中的分区信息,可以看到 city=beijing 已经被删除:

```
0: jdbc:hive2://hadoop3:10000 > show partitions student_ptn;
OK
city=HK
city=chongqing
city=guangzhou
city=nanjing
city=shanghai
Time taken: 0.971 seconds, Fetched: 5 row(s)
```

4. 删除表

0：jdbc：hive2：//hadoop3：10000 > drop table new_student；

5. 清空表

0：jdbc：hive2：//hadoop3：10000 > truncate table student_ptn；

6. Hive 常用命令

Hive 常用命令如表 5-4 所示。

表 5-4 Hive 常用命令

命令	描述
show databases	查看数据库列表
show tables	查看数据表
show create table table_name	查看数据表的建表命令
show functions	查看 Hive 函数
show partition table_name	查看 Hive 表的分区
desc table_name desc extended table_name desc formatted table_name	查看表的详细信息
desc database db_name desc database extended db_name	查看数据仓库的详细属性信息
truncate table table_name	清空表
drop table table_name	删除表

5.5 Hive 应用案例

5.5.1 统计单月访问次数和总访问次数

本案例模拟统计电商用户单月访问次数和总访问次数，实现基本的用户访问统计画像。

1. 数据说明

数据存储在/home/hadooop/user_access.txt 文件中。数据字段说明：用户名，月份，访问次数。

文件内容显示如下：

```
A,2020-01,5
A,2020-01,15
B,2020-01,5
A,2020-01,8
B,2020-01,25
```

```
A,2020-01,5
A,2020-02,4
A,2020-02,6
B,2020-02,10
B,2020-02,5
A,2020-03,16
A,2020-03,22
B,2020-03,23
B,2020-03,10
B,2020-03,1
```

2. 数据准备

(1) 创建表

hive> use t2;
hive> create external table if not exists user_access(
uname string comment '用户名',
umonth string comment '月份',
ucount int comment '访问次数'
) comment '用户访问表'
row format delimited fields terminated by ","
location "/hive/user_access";

(2) 导入数据

hive> load data local inpath "/home/hadoop/user_access.txt" into table user_access;

(3) 验证数据

hive> select * from user_access;

3. 结果需求

需求分析：统计每个用户截止到每月底的最大单月访问次数和累计到该月的总访问次数。

4. 需求分析

此结果需要根据用户名＋月份进行分组。

(1) 先求出当月访问次数

//求当月访问次数
hive> create table tmp_access(
name string,
mon string,
num int
);
//插入数据
hive> insert into table tmp_access select uname,umonth,sum(ucount) from user_access t group by t.uname,t.umonth;

屏幕提示如下信息：

```
    WARNING: Hive-on-MR is deprecated in Hive 2 and may not be available in the future versions.
Consider using a different execution engine (i.e. spark, tez) or using Hive 1.X releases.
    Query ID = hadoop_20200428130324_41cf68bb-3424-4fd9-bc07-51edb1c994a2
    Total jobs = 1
    Launching Job 1 out of 1
    Number of reduce tasks not specified. Estimated from input data size: 1
    In order to change the average load for a reducer (in bytes):
      set hive.exec.reducers.bytes.per.reducer=<number>
    In order to limit the maximum number of reducers:
      set hive.exec.reducers.max=<number>
    In order to set a constant number of reducers:
      set mapreduce.job.reduces=<number>
    Job running in-process (local Hadoop)
    2020-04-28 13:03:34,224 Stage-1 map = 0%,  reduce = 0%
    2020-04-28 13:03:40,727 Stage-1 map = 100%, reduce = 0%
    2020-04-28 13:03:42,881 Stage-1 map = 100%, reduce = 100%
    Ended Job = job_local57430813_0001
    Loading data to table t2.tmp_access
    MapReduce Jobs Launched:
    Stage-Stage-1:  HDFS Read: 1760 HDFS Write: 1538 SUCCESS
    Total MapReduce CPU Time Spent: 0 msec
    OK
    Time taken: 19.697 seconds
```

//查询临时数据

hive> select * from tmp_access;

查询结果参考如下输出:

```
OK
    NULL    NULL
A   2020-01 33
A   2020-02 10
A   2020-03 38
B   2020-01 30
B   2020-02 15
B   2020-03 34
Time taken: 0.679 seconds, Fetched: 7 row(s)
```

(2) tmp_access 进行自连接视图

hive> create view tmp_view as select a.name anme,a.mon amon,a.num anum,b.name bname,b.mon bmon,b.num bnum from tmp_access a join tmp_access b on a.name = b.name;
hive> select * from tmp_view;

查询结果参考如下输出:

```
Execution completed successfully
MapredLocal task succeeded
Launching Job 1 out of 1
```

```
Number of reduce tasks is set to 0 since there's no reduce operator
Job running in-process (local Hadoop)
2020-04-28 13:10:23,062 Stage-3 map = 100%, reduce = 0%
Ended Job = job_local106541202_0002
MapReduce Jobs Launched:
Stage-Stage-3: HDFS Read: 1119 HDFS Write: 846 SUCCESS
Total MapReduce CPU Time Spent: 0 msec
OK
        NULL    NULL            NULL    NULL
A       2020-01 33      A       2020-01 33
A       2020-02 10      A       2020-01 33
A       2020-03 38      A       2020-01 33
A       2020-01 33      A       2020-02 10
A       2020-02 10      A       2020-02 10
A       2020-03 38      A       2020-02 10
A       2020-01 33      A       2020-03 38
A       2020-02 10      A       2020-03 38
A       2020-03 38      A       2020-03 38
B       2020-01 30      B       2020-01 30
B       2020-02 15      B       2020-01 30
B       2020-03 34      B       2020-01 30
B       2020-01 30      B       2020-02 15
B       2020-02 15      B       2020-02 15
B       2020-03 34      B       2020-02 15
B       2020-01 30      B       2020-03 34
B       2020-02 15      B       2020-03 34
B       2020-03 34      B       2020-03 34
Time taken: 92.876 seconds, Fetched: 19 row(s)
```

(3) 进行比较统计

hive> select anme,amon,anum,max(bnum) as max_access,sum(bnum) as sum_access from tmp_view where amon>=bmon group by anme,amon,anum;

查询结果参考如下输出：

```
Execution completed successfully
MapredLocal task succeeded
Launching Job 1 out of 1
Number of reduce tasks not specified. Estimated from input data size: 1
In order to change the average load for a reducer (in bytes):
    set hive.exec.reducers.bytes.per.reducer=<number>
In order to limit the maximum number of reducers:
    set hive.exec.reducers.max=<number>
In order to set a constant number of reducers:
    set mapreduce.job.reduces=<number>
Job running in-process (local Hadoop)
2020-04-28 13:33:09,641 Stage-2 map = 0%, reduce = 0%
```

```
2020-04-28 13:33:10,686 Stage-2 map = 100%,  reduce = 100%
Ended Job = job_local2047342435_0003
MapReduce Jobs Launched:
Stage-Stage-2:  HDFS Read: 2408 HDFS Write: 1692 SUCCESS
Total MapReduce CPU Time Spent: 0 msec
OK
A       2020-01 33      33      33
A       2020-02 10      33      43
A       2020-03 38      38      81
B       2020-01 30      30      30
B       2020-02 15      30      45
B       2020-03 34      34      79
Time taken: 69.045 seconds, Fetched: 6 row(s)
```

5.5.2　学生课程成绩统计

课程表 course.txt 内容如下：

```
1,数据库
2,数学
3,信息系统
4,操作系统
5,数据结构
6,数据处理
```

学生表 student.txt 内容如下：

```
95001,李勇,男,20,CS
95002,刘晨,女,19,IS
95003,王敏,女,22,MA
95004,张立,男,19,IS
95005,刘刚,男,18,MA
95006,孙庆,男,23,CS
95007,易思玲,女,19,MA
95008,李娜,女,18,CS
95009,梦圆圆,女,18,MA
95010,孔小涛,男,19,CS
95011,包小柏,男,18,MA
95012,孙花,女,20,CS
95013,冯伟,男,21,CS
95014,王小丽,女,19,CS
95015,王君,男,18,MA
95016,钱国,男,21,MA
95017,王风娟,女,18,IS
95018,王一,女,19,IS
95019,邢小丽,女,19,IS
95020,赵钱,男,21,IS
```

95021,周二,男,17,MA
95022,郑明,男,20,MA

学生成绩表 score.txt 内容如下：

95001,1,81
95001,2,85
95001,3,88
95001,4,70
95002,2,90
95002,3,80
95002,4,71
95002,5,60
95003,1,82
95003,3,90
95003,5,100
95004,1,80
95004,2,92
95004,4,91
95004,5,70
95005,1,70
95005,2,92
95005,3,99
95005,6,87
95006,1,72
95006,2,62
95006,3,100
95006,4,59
95006,5,60
95006,6,98
95007,3,68
95007,4,91
95007,5,94
95007,6,78
95008,1,98
95008,3,89
95008,6,91
95009,2,81
95009,4,89
95009,6,100
95010,2,98
95010,5,90
95010,6,80
95011,1,81
95011,2,91
95011,3,81
95011,4,86

```
95012,1,81
95012,3,78
95012,4,85
95012,6,98
95013,1,98
95013,2,58
95013,4,88
95013,5,93
95014,1,91
95014,2,100
95014,4,98
95015,1,91
95015,3,59
95015,4,100
95015,6,95
95016,1,92
95016,2,99
95016,4,82
95017,4,82
95017,5,100
95017,6,58
95018,1,95
95018,2,100
95018,3,67
95018,4,78
95019,1,77
95019,2,90
95019,3,91
95019,4,67
95019,5,87
95020,1,66
95020,2,99
95020,5,93
95021,2,93
95021,5,91
95021,6,99
95022,3,69
95022,4,93
95022,5,82
95022,6,100
```

1. 创建表以及录入数据

```
hive> create external table if not exists course(
courseid string,
coursename string
)
```

```
row format delimited fields terminated by ','
stored as textfile location '/hive/course';
hive> load data local inpath '/home/hadoop/course.txt' into table course;

hive> create external table if not exists score(
userid string,
courseid string,
score string
)
row format delimited fields terminated by ','
stored as textfile location '/hive/score';
hive> load data local inpath '/home/hadoop/score.txt' into table score;

hive> create external table if not exists student(
userid string,
name string,
sex string,
age string,
xi string
)
row format delimited fields terminated by ','
stored as textfile location '/hive/student';
hive> load data local inpath '/home/hadoop/student.txt' into table student;
```

2. 需求查询

(1) 问题:查询全体学生的学号与姓名

```
hive> select userid,name from student;
```

输出结果参考如下:

```
OK
95002    刘晨
95017    王风娟
95018    王一
95013    冯伟
95014    王小丽
95019    邢小丽
95020    赵钱
95003    王敏
95004    张立
95012    孙花
95010    孔小涛
95005    刘刚
95006    孙庆
95007    易思玲
95008    李娜
95021    周二
```

95022	郑明
95001	李勇
95011	包小柏
95009	梦圆圆
95015	王君

Time taken: 0.416 seconds, Fetched: 21 row(s)

(2) 问题:查询选修了课程的学生姓名和课程名称

hive> select student.name,t.coursename
from(
select course.coursename coursename,score.userid userid
from score score,course course
where score.courseid = course.courseid) t,student
where t.userid = student.userid;

输出结果参考如下:

Execution completed successfully
MapredLocal task succeeded
Launching Job 1 out of 1
Number of reduce tasks is set to 0 since there's no reduce operator
Job running in-process (local Hadoop)
2020-04-28 14:39:31,466 Stage-5 map = 0%, reduce = 0%
2020-04-28 14:39:36,691 Stage-5 map = 100%, reduce = 0%
Ended Job = job_local684754266_0001
MapReduce Jobs Launched:
Stage-Stage-5: HDFS Read: 910 HDFS Write: 0 SUCCESS
Total MapReduce CPU Time Spent: 0 msec
OK

李勇	数据库
李勇	数学
李勇	信息系统
李勇	操作系统
刘晨	数学
刘晨	信息系统
刘晨	操作系统
刘晨	数据结构
王敏	数据库
王敏	信息系统
王敏	数据结构
张立	数据库
张立	数学
张立	操作系统
张立	数据结构
刘刚	数据库
刘刚	数学
刘刚	信息系统

刘刚	数据处理
孙庆	数据库
孙庆	数学
孙庆	信息系统
孙庆	操作系统
孙庆	数据结构
孙庆	数据处理
易思玲	信息系统
易思玲	操作系统
易思玲	数据结构
易思玲	数据处理
李娜	数据库
李娜	信息系统
李娜	数据处理
梦圆圆	数学
梦圆圆	操作系统
梦圆圆	数据处理
孔小涛	数学
孔小涛	数据结构
孔小涛	数据处理
包小柏	数据库
包小柏	数学
包小柏	信息系统
包小柏	操作系统
孙花	数据库
孙花	信息系统
孙花	操作系统
孙花	数据处理
冯伟	数据库
冯伟	数学
冯伟	操作系统
冯伟	数据结构
王小丽	数据库
王小丽	数学
王小丽	操作系统
王君	数据库
王君	信息系统
王君	操作系统
王君	数据处理
王凤娟	操作系统
王凤娟	数据结构
王凤娟	数据处理
王一	数据库
王一	数学
王一	信息系统
王一	操作系统

邢小丽	数据库
邢小丽	数学
邢小丽	信息系统
邢小丽	操作系统
邢小丽	数据结构
赵钱	数据库
赵钱	数学
赵钱	数据结构
周二	数学
周二	数据结构
周二	数据处理
郑明	信息系统
郑明	操作系统
郑明	数据结构
郑明	数据处理

Time taken: 86.075 seconds, Fetched: 79 row(s)

（3）问题：查询每个选修课程共选了多少人

```
hive> select t.coursename,count( * ) num from(
select course.coursename coursename,scor.userid userid
from score,course
where score.courseid = course.courseid) t,student
where t.userid = student.userid
group by t.coursename
order by num desc
limit 3;
```

输出结果参考如下：

```
Execution completed successfully
MapredLocal task succeeded
Launching Job 1 out of 2
Number of reduce tasks not specified. Estimated from input data size: 1
In order to change the average load for a reducer (in bytes):
    set hive.exec.reducers.bytes.per.reducer=<number>
In order to limit the maximum number of reducers:
    set hive.exec.reducers.max=<number>
In order to set a constant number of reducers:
    set mapreduce.job.reduces=<number>
Job running in-process (local Hadoop)
2020-04-28 14:46:17,533 Stage-3 map = 0%,  reduce = 0%
2020-04-28 14:46:18,622 Stage-3 map = 100%, reduce = 0%
2020-04-28 14:46:19,714 Stage-3 map = 100%, reduce = 100%
Ended Job = job_local261505760_0002
Launching Job 2 out of 2
Number of reduce tasks determined at compile time: 1
In order to change the average load for a reducer (in bytes):
```

```
    set hive.exec.reducers.bytes.per.reducer=<number>
In order to limit the maximum number of reducers:
    set hive.exec.reducers.max=<number>
In order to set a constant number of reducers:
    set mapreduce.job.reduces=<number>
Job running in-process (local Hadoop)
2020-04-28 14:46:23,576 Stage-4 map = 100%, reduce = 100%
Ended Job = job_local1678226515_0003
MapReduce Jobs Launched:
Stage-Stage-3: HDFS Read: 3640 HDFS Write: 0 SUCCESS
Stage-Stage-4: HDFS Read: 3640 HDFS Write: 0 SUCCESS
Total MapReduce CPU Time Spent: 0 msec
OK
操作系统        15
数学    14
数据库    14
Time taken: 68.595 seconds, Fetched: 3 row(s)
```

(4) 问题:查询学生的总人数

hive> select count(distinct(userid)) from student;

输出结果参考如下:

```
    WARNING: Hive-on-MR is deprecated in Hive 2 and may not be available in the future versions.
Consider using a different execution engine (i.e. spark, tez) or using Hive 1.X releases.
    Query ID = hadoop_20200428144922_41ec9dfe-50df-4a74-adc9-6018482fb370
Total jobs = 1
Launching Job 1 out of 1
Number of reduce tasks determined at compile time: 1
In order to change the average load for a reducer (in bytes):
    set hive.exec.reducers.bytes.per.reducer=<number>
In order to limit the maximum number of reducers:
    set hive.exec.reducers.max=<number>
In order to set a constant number of reducers:
    set mapreduce.job.reduces=<number>
Job running in-process (local Hadoop)
2020-04-28 14:49:28,032 Stage-1 map = 0%, reduce = 0%
2020-04-28 14:49:29,135 Stage-1 map = 100%, reduce = 100%
Ended Job = job_local1768135327_0004
MapReduce Jobs Launched:
Stage-Stage-1: HDFS Read: 4648 HDFS Write: 0 SUCCESS
Total MapReduce CPU Time Spent: 0 msec
OK
21
Time taken: 6.337 seconds, Fetched: 1 row(s)
```

(5) 问题:计算数据库课程的学生平均成绩

hive > select course.coursename,avg(score.score)
from score,course
where course.coursename = '数据库' and score.courseid = course.courseid
group by course.coursename;

输出结果参考如下:

```
Execution completed successfully
MapredLocal task succeeded
Launching Job 1 out of 1
Number of reduce tasks not specified. Estimated from input data size: 1
In order to change the average load for a reducer (in bytes):
    set hive.exec.reducers.bytes.per.reducer = <number>
In order to limit the maximum number of reducers:
    set hive.exec.reducers.max = <number>
In order to set a constant number of reducers:
    set mapreduce.job.reduces = <number>
Job running in-process (local Hadoop)
2020-04-28 15:01:38,876 Stage-2 map = 0%,  reduce = 0%
2020-04-28 15:01:39,909 Stage-2 map = 100%,  reduce = 100%
Ended Job = job_local2058382398_0005
MapReduce Jobs Launched:
Stage-Stage-2:  HDFS Read: 6468 HDFS Write: 0 SUCCESS
Total MapReduce CPU Time Spent: 0 msec
OK
数据库    83.66666666666667
Time taken: 84.829 seconds, Fetched: 1 row(s)
```

(6) 问题:查询选修数学课程的学生的最高分数

hive > select max(score.score)
from score,course
where course.coursename = '数学' and score.courseid = course.courseid;

输出结果参考如下:

```
Execution completed successfully
MapredLocal task succeeded
Launching Job 1 out of 1
Number of reduce tasks determined at compile time: 1
In order to change the average load for a reducer (in bytes):
    set hive.exec.reducers.bytes.per.reducer = <number>
In order to limit the maximum number of reducers:
    set hive.exec.reducers.max = <number>
In order to set a constant number of reducers:
    set mapreduce.job.reduces = <number>
Job running in-process (local Hadoop)
2020-04-28 15:12:59,429 Stage-2 map = 0%,  reduce = 0%
2020-04-28 15:13:00,468 Stage-2 map = 100%,  reduce = 100%
```

Ended Job = job_local1308249922_0006
MapReduce Jobs Launched:
Stage-Stage-2: HDFS Read: 8288 HDFS Write: 0 SUCCESS
Total MapReduce CPU Time Spent: 0 msec
OK
99
Time taken: 70.314 seconds, Fetched: 1 row(s)

(7) 问题:查询选修了 3 门以上课程的学生姓名

hive> select student.name,t.num from(
select userid,count(*) num
from score
group by userid) t, student
where t.userid = student.userid and t.num >= 3;

输出结果参考如下:

```
Execution completed successfully
MapredLocal task succeeded
Launching Job 2 out of 2
Number of reduce tasks is set to 0 since there's no reduce operator
Job running in-process (local Hadoop)
2020-04-28 15:15:35,317 Stage-4 map = 100%, reduce = 0%
Ended Job = job_local387073030_0008
MapReduce Jobs Launched:
Stage-Stage-1: HDFS Read: 10108 HDFS Write: 0 SUCCESS
Stage-Stage-4: HDFS Read: 5054 HDFS Write: 0 SUCCESS
Total MapReduce CPU Time Spent: 0 msec
OK
李勇     4
刘晨     4
王敏     3
张立     4
刘刚     4
孙庆     6
易思玲   4
李娜     3
梦圆圆   3
孔小涛   3
包小柏   4
孙花     4
冯伟     4
王小丽   3
王君     4
王风娟   3
王一     4
邢小丽   5
```

赵钱	3
周二	3
郑明	4

Time taken: 59.135 seconds, Fetched: 21 row(s)

(8) 问题：按照年龄排序并直接输出到不同的文件中

```
hive> create table if not exists result2(
userid string,
name string,
sex string,
age string,
xi string
)
row format delimited fields terminated by ',' 
stored as textfile;
hive> insert into result2 select * from student order by age desc;
```

输出结果参考如下：

```
    WARNING: Hive-on-MR is deprecated in Hive 2 and may not be available in the future versions.
Consider using a different execution engine (i.e. spark, tez) or using Hive 1.X releases.
    Query ID = hadoop_20200428151913_4daf459c-98bb-4117-8896-7d8a6b140ad9
    Total jobs = 1
    Launching Job 1 out of 1
    Number of reduce tasks determined at compile time: 1
    In order to change the average load for a reducer (in bytes):
        set hive.exec.reducers.bytes.per.reducer=<number>
    In order to limit the maximum number of reducers:
        set hive.exec.reducers.max=<number>
    In order to set a constant number of reducers:
        set mapreduce.job.reduces=<number>
    Job running in-process (local Hadoop)
    2020-04-28 15:19:17,987 Stage-1 map = 100%,  reduce = 0%
    2020-04-28 15:19:20,046 Stage-1 map = 100%,  reduce = 100%
    Ended Job = job_local183767599_0009
    Loading data to table t1.result2
    MapReduce Jobs Launched:
    Stage-Stage-1:  HDFS Read: 11116 HDFS Write: 571 SUCCESS
    Total MapReduce CPU Time Spent: 0 msec
    OK
    Time taken: 8.47 seconds
```

(9) 问题：查询学生的得分情况

```
hive> select student.name,score.score from score,student
where score.userid=student.userid;
```

输出结果参考如下：

```
Execution completed successfully
MapredLocal task succeeded
Launching Job 1 out of 1
Number of reduce tasks is set to 0 since there's no reduce operator
Job running in-process (local Hadoop)
2020-04-28 15:23:13,186 Stage-3 map = 100%, reduce = 0%
Ended Job = job_local459741515_0010
MapReduce Jobs Launched:
Stage-Stage-3: HDFS Read: 6535 HDFS Write: 571 SUCCESS
Total MapReduce CPU Time Spent: 0 msec
OK
李勇     81
李勇     85
李勇     88
李勇     70
刘晨     90
刘晨     80
刘晨     71
刘晨     60
王敏     82
王敏     90
王敏     100
张立     80
张立     92
张立     91
张立     70
刘刚     70
刘刚     92
刘刚     99
刘刚     87
孙庆     72
孙庆     62
孙庆     100
孙庆     59
孙庆     60
孙庆     98
易思玲   68
易思玲   91
易思玲   94
易思玲   78
李娜     98
李娜     89
李娜     91
梦圆圆   81
梦圆圆   89
梦圆圆   100
```

孔小涛	98
孔小涛	90
孔小涛	80
包小柏	81
包小柏	91
包小柏	81
包小柏	86
孙花	81
孙花	78
孙花	85
孙花	98
冯伟	98
冯伟	58
冯伟	88
冯伟	93
王小丽	91
王小丽	100
王小丽	98
王君	91
王君	59
王君	100
王君	95
王风娟	82
王风娟	100
王风娟	58
王一	95
王一	100
王一	67
王一	78
邢小丽	77
邢小丽	90
邢小丽	91
邢小丽	67
邢小丽	87
赵钱	66
赵钱	99
赵钱	93
周二	93
周二	91
周二	99
郑明	69
郑明	93
郑明	82
郑明	100

Time taken: 61.814 seconds, Fetched: 79 row(s)

（10）问题：查询选修信息系统课程且成绩在 90 分以上的所有学生

hive > select distinct(t2.name) from(
select score.userid userid
from score,course
where score.score > = 90 and score.courseid = course.courseid) t1,student t2
where t1.userid = t2.userid;

输出结果参考如下：

```
Execution completed successfully
MapredLocal task succeeded
Launching Job 1 out of 1
Number of reduce tasks not specified. Estimated from input data size：1
In order to change the average load for a reducer（in bytes）：
    set hive.exec.reducers.bytes.per.reducer = < number >
In order to limit the maximum number of reducers：
    set hive.exec.reducers.max = < number >
In order to set a constant number of reducers：
    set mapreduce.job.reduces = < number >
Job running in-process（local Hadoop）
2020-04-28 15：26：33,716 Stage-3 map = 0％, reduce = 0％
2020-04-28 15：26：34,751 Stage-3 map = 100％, reduce = 100％
Ended Job = job_local1046991012_0011
MapReduce Jobs Launched：
Stage-Stage-3: HDFS Read：14890 HDFS Write：1142 SUCCESS
Total MapReduce CPU Time Spent：0 msec
OK
冯伟
刘刚
刘晨
包小柏
周二
孔小涛
孙庆
孙花
张立
易思玲
李娜
梦圆圆
王一
王君
王小丽
王敏
王凤娟
赵钱
邢小丽
```

郑明
Time taken: 67.653 seconds, Fetched: 20 row(s)

(11) 问题:查询与"刘晨"在同一个系学习的其他学生

hive> select t3.name from(
select t2.name name
from student t1,student t2
where t1.name='刘晨' and t1.xi = t2.xi) t3
where t3.name<>'刘晨';

输出结果参考如下:

```
Execution completed successfully
MapredLocal task succeeded
Launching Job 1 out of 1
Number of reduce tasks is set to 0 since there's no reduce operator
Job running in-process (local Hadoop)
2020-04-28 15:29:11,230 Stage-3 map = 100%, reduce = 0%
Ended Job = job_local1211379364_0012
MapReduce Jobs Launched:
Stage-Stage-3:  HDFS Read: 7949 HDFS Write: 571 SUCCESS
Total MapReduce CPU Time Spent: 0 msec
OK
王凤娟
王一
邢小丽
赵钱
张立
Time taken: 64.538 seconds, Fetched: 5 row(s)
```

本 章 小 结

① Hive 是一个基于 Hadoop 生态的分布式数据仓库工具,主要用来在分布式系统 Hadoop 框架上进行数据分析,其目的是分析查询结构化的海量数据。实际上 Hive 最早是 Facebook 用于分析大量的数据用户和日志数据的。我们之前学习的 Hadoop 分布式文件系统、MapReduce 并行计算框架、HBase 分布式数据库在程序中使用 API 基本都涉及 Java 编程,所以需要掌握一定的 Java 编程能力。Hive 正是给那些只懂得 SQL 语句,而不太懂得 Java 编程的人使用的。Hive 提供 HQL(类 SQL 编程接口),能直接转化为 MapReduce 程序在 Hadoop 平台上运行。

② Hive 数据仓库主要有 4 个数据模型:内部表(Table)、外部表(External Table)、分区(Partition)和桶(Bucket)。

③ Hive 系统安装自带 Derby 数据库,用于 Hive 元数据存储。但为了让多用户和远程可访问,生产场景中一般更换成更大的数据库(如 MySQL)来实现 Hive 元数据存储。

④ Hive 不适合用于联机(online)事务处理，也不提供实时查询功能，它最适合应用于基于大量不可变数据的批处理作业。Hive 的特点包括：可伸缩（在 Hadoop 的集群上动态添加设备）、可扩展、容错、输入格式的松散耦合。

第 6 章 Kafka 消息系统

Kafka 是一个分布式的，基于发布、订阅的消息系统，具有高吞吐、高容错、高可靠以及高性能等特性，主要用于应用解耦、流量削峰、异步消息等场景。

为了让大家更加深入地了解 Kafka 消息中间件内部实现原理，后续内容中将会从主题与日志开始介绍消息的存储、删除以及检索，然后介绍其副本机制的实现原理，最后介绍生产与消费的实现原理，以便更合理地应用于实际业务。

6.1 Kafka 消息系统的功能

6.1.1 Kafka 概述

1. 简介

Kafka 是一个分布式的，基于发布、订阅的消息系统，有着强大的消息处理能力，相比于其他消息系统，具有以下特性。

① 快速数据持久化，实现了 $O(1)$ 时间复杂度的数据持久化能力。

② 高吞吐，能在普通的服务器上达到 10 万次消息处理每秒的吞吐速率。

③ 高可靠，消息持久化以及副本系统的机制保证了消息的可靠性，消息可以多次消费。

④ 高扩展，与其他分布式系统一样，所有组件均支持分布式、自动实现负载均衡，可以快速便捷地扩容系统。

⑤ 离线与实时处理能力并存，提供了在线与离线的消息处理能力。

Kafka 凭借这些优秀特性而广泛用于应用解耦、流量削峰、异步消息等场景，如消息中间件、日志聚合、流处理等应用。

2. 消息系统

一个消息系统负责将数据从一个应用传递到另外一个应用，应用只需关注数据，无须关注数据在两个或多个应用间是如何传递的。分布式消息传递基于可靠的消息队列，在客户端应用和消息系统之间异步传递消息。有两种主要的消息传递模式：点对点传递模式、发布-订阅模式。大部分的消息系统选用发布-订阅模式。Kafka 就是一种发布-订阅模式。

（1）点对点消息系统

在点对点消息系统中，消息被持久化到一个队列（Queue）中。此时，将有一个或多个消费者消费队列中的数据。但是一条消息只能被消费一次,当一个消费者消费了队列中的某条数据之后,该条数据则从消息队列中删除。该模式中即使有多个消费者同时消费数据,也能保证数据处理的顺序。这种模式的示意图如图 6-1 所示。

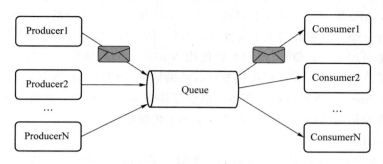

图 6-1 点对点消息系统示意图

(2)发布-订阅消息系统

在发布-订阅消息系统中,消息被持久化到一个 Topic(主题)中。与点对点消息系统不同的是,消费者可以订阅一个或多个 Topic,消费者可以消费该 Topic 中所有的数据,同一条数据可以被多个消费者消费,数据被消费后不会立刻删除。在发布-订阅消息系统中,消息的生产者称为发布者,消费者称为订阅者。该模式的示意图如图 6-2 所示。

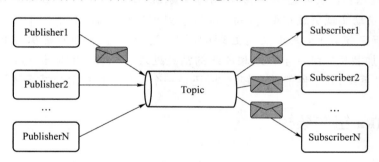

图 6-2 发布-订阅消息系统示意图

3. 常用 Message Queue 对比

(1)RabbitMQ

RabbitMQ 是使用 Erlang 编写的一个开源的消息队列,本身支持很多的协议,如 AMQP、XMPP、SMTP、STOMP,相对来说比较重量级,更适用于企业级的开发。同时实现了 Broker 构架,这意味着消息在发送给客户端时先在中心队列排队。对路由、负载均衡或者数据持久化都有很好的支持。

(2)Redis

Redis 是一个基于 key-value 对的 NoSQL 数据库,开发维护很活跃。虽然它是一个 key-value 数据库存储系统,但它本身支持 MQ 功能,所以完全可以当作一个轻量级的队列服务来使用。对于 RabbitMQ 和 Redis 的入队和出队操作,各执行 100 万次,每 10 万次记录一次执行时间。测试数据分为 128 B、512 B、1 KB 和 10 KB 4 个不同大小的数据。实验表明:入队时,当数据比较小时 Redis 的性能要高于 RabbitMQ,而如果数据大小超过了 10 KB,Redis 则慢得无法忍受;出队时,无论数据大小,Redis 都表现出非常好的性能,而 RabbitMQ 的出队性能远低于 Redis。

(3) ZeroMQ

ZeroMQ 号称最快的消息队列系统,尤其针对大吞吐量的需求场景。ZeroMQ 能够实现 RabbitMQ 不擅长的高级/复杂的队列,但是开发人员需要自己组合多种技术框架,技术上的复杂度是对 ZeroMQ 能够应用成功的挑战。ZeroMQ 具有一个独特的非中间件模式,用户不需要安装和运行一个消息服务器或中间件,因为用户的应用程序将扮演这个服务器角色。用户只需要简单地引用 ZeroMQ 程序库(可以使用 NuGet 安装),然后就可以愉快地在应用程序之间发送消息了。但是 ZeroMQ 仅提供非持久性的队列,也就是说,如果宕机,则数据将会丢失。Twitter 的 Storm 0.9.0 以前的版本中默认使用 ZeroMQ 作为数据流的传输模块(Storm 从 0.9.0 版本开始同时支持 ZeroMQ 和 Netty 作为传输模块)。

(4) ActiveMQ

ActiveMQ 是 Apache 下的一个子项目。类似于 ZeroMQ,它能够以代理人和点对点的技术实现队列。同时类似于 RabbitMQ,它用少量代码就可以高效地实现高级应用场景。

(5) Kafka/Jafka

Kafka 是 Apache 下的一个子项目,是一个高性能、跨语言、分布式发布-订阅消息队列系统,而 Jafka 是在 Kafka 之上孵化而来的,即 Kafka 的一个升级版。Kafka 具有以下特性:快速持久化,可以在 $O(1)$ 的系统开销下进行消息持久化;高吞吐,在一台普通的服务器上即可达到 10 万次消息处理每秒的吞吐速率;完全的分布式系统,Broker、Producer、Consumer 都原生自动支持分布式,自动实现负载均衡;支持 Hadoop 数据并行加载,对于像 Hadoop 一样的日志数据和离线分析系统,但又要求实时处理的限制,这是一个可行的解决方案。Kafka 通过 Hadoop 的并行加载机制统一了在线和离线的消息处理。Kafka 相对于 ActiveMQ 是一个非常轻量级的消息系统,除了性能非常好之外,还是一个工作良好的分布式系统。

6.1.2 Kafka 组件架构

1. Broker

Kafka 集群包含一个或多个服务器,服务器节点称为 Broker,Broker 存储 Topic 的数据。如果某 Topic 有 N 个 Partition,集群有 N 个 Broker,那么每个 Broker 存储该 Topic 的一个 Partition。

2. Topic

每条发布到 Kafka 集群的消息都有一个类别,这个类别称为 Topic,类似于数据库的表名。

3. Partition

一个 Partition 中的数据使用多个 Segment 文件存储。Partition 中的数据是有序的,不同 Partition 间的数据丢失了数据的顺序。如果 Topic 有多个 Partition,消费数据时就不能保证数据的顺序。在需要严格保证消息的消费顺序的场景下,需要将 Partition 数目设为 1。

4. Producer

Producer(生产者)即数据的发布者,该角色将消息发布到 Kafka 的 Topic 中。Broker 接收到生产者发送的消息后,将该消息追加到当前用于追加数据的 Segment 文件中。生产者发送的消息存储到一个 Partition 中,生产者也可以指定存储数据的 Partition。

5. Consumer

Consumer(消费者)可以从 Broker 中读取数据。消费者可以消费多个 Topic 中的数据。

6. Consumer Group

每个 Consumer 属于一个特定的 Consumer Group(可为每个 Consumer 指定 group name,若不指定 group name 则属于默认的 Group)。

7. Leader

每个 Partition 有多个副本,其中有且仅有一个作为 Leader,Leader 是当前负责数据读写的 Partition。

8. Follower

Follower 跟随 Leader,所有写请求都通过 Leader 路由,数据变更会广播给所有 Follower,Follower 与 Leader 保持数据同步。如果 Leader 失效,则从 Follower 中选举出一个新的 Leader。当 Follower 与 Leader 挂掉、卡住或者同步太慢时,Leader 会把这个 Follower 从 In-Sync Replicas(ISR,副本同步队列)列表中删除,重新创建一个 Follower。

Kafka 组件之间的关系以及整体架构可以参考图 6-3。

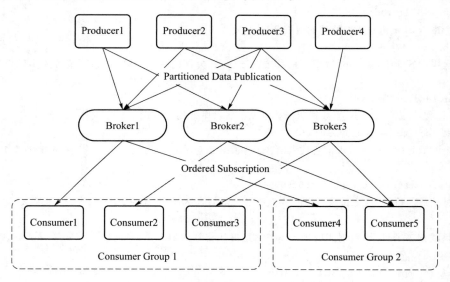

图 6-3 Kafka 组件之间的关系以及整体架构

6.1.3 Kafka 软件安装

1. Kafka 软件下载

本章实验环境中,我们使用的版本为 kafka_2.11-0.8.2.0.tgz,下载网址可以访问 Kafka 官网,建议使用国内镜像站点 http://mirrors.hust.edu.cn/apache/,以获得更好的下载体验。

2. 上传解压缩

```
[hadoop@hadoop1 ~]$ tar -zxvf kafka_2.11-0.8.2.0.tgz -C apps
[hadoop@hadoop1 ~]$ cd apps/
[hadoop@hadoop1 apps]$ ln -s kafka_2.11-0.8.2.0/ kafka
```

3. 修改配置文件

进入 Kafka 的安装配置目录：

```
[hadoop@hadoop1 ~]$ cd apps/kafka/config/
[hadoop@hadoop1 ~]$ vim server.properties
```

编辑 server.properties 文件即可。我们可以发现在目录下有很多文件，如 zookeeper 文件，我们可以根据 Kafka 内带的 ZK 集群来启动，但是建议使用独立的 ZK 集群。值得注意的是 server.properties（每个节点的 broker.id 和 host.name 都不相同）。

```
//当前机器在集群中的唯一标识,和 ZooKeeper 的 myid 性质一样
broker.id = 0
//当前 Kafka 对外提供服务的端口默认是 9092
port = 9092
//这个参数默认是关闭的,在 0.8.1 版本中有个 bug:DNS 解析问题失败率的问题
host.name = hadoop1
//这个是 Broker 进行网络处理的线程数
num.network.threads = 3
//这个是 Broker 进行 I/O 处理的线程数
num.io.threads = 8
//发送缓冲区 Buffer 大小,数据不是直接发送的,先会存储到缓冲区,达到一定的大小后再发送,能提高性能
socket.send.buffer.bytes = 102400
//Kafka 接收缓冲区大小,当数据达到一定大小后再序列化到磁盘
socket.receive.buffer.bytes = 102400
//这个参数是向 Kafka 请求消息或者向 Kafka 发送消息的请求的最大数,这个值不能超过 Java 的堆栈大小
socket.request.max.bytes = 104857600
//消息存放的目录,这个目录可以配置为逗号分割的表达式,上面的 num.io.threads 要大于这个目录的个数
//如果配置多个目录,新创建的 Topic 把消息持久化到的地方是,当前以逗号分割的目录中,哪个分区数最少就放到哪一个中
log.dirs = /home/hadoop/log/kafka-logs
//默认的分区数,一个 Topic 默认有 1 个分区
num.partitions = 1
//每个数据目录用于日志恢复的线程数
num.recovery.threads.per.data.dir = 1
//默认消息的最大持久化时间,168 小时,7 天
log.retention.hours = 168
//这个参数是:因为 Kafka 的消息是以追加的形式落地到文件,当超过这个值的时候,Kafka 会新起一个文件
log.segment.bytes = 1073741824
//每隔 300 000 毫秒去检查上面配置的 Log 失效时间
log.retention.check.interval.ms = 300000
//是否启用 Log 压缩,一般不启用,启用的话可以提高性能
log.cleaner.enable = false
//设置 ZooKeeper 的连接端口
```

```
zookeeper.connect=192.168.123.102:2181,192.168.123.103:2181,192.168.123.104:2181
//设置 ZooKeeper 的连接超时时间
zookeeper.connection.timeout.ms=6000
producer.properties
metadata.broker.list=192.168.123.102:9092,192.168.123.103:9092,192.168.123.104:9092
consumer.properties
zookeeper.connect=192.168.123.102:2181,192.168.123.103:2181,192.168.123.104:2181
```

4．将 Kafka 的安装包分发到其他节点

[hadoop@hadoop1 apps]$ scp -r kafka_2.11-0.8.2.0/ hadoop2:$PWD
[hadoop@hadoop1 apps]$ scp -r kafka_2.11-0.8.2.0/ hadoop3:$PWD
[hadoop@hadoop1 apps]$ scp -r kafka_2.11-0.8.2.0/ hadoop4:$PWD

5．创建软连接

[hadoop@hadoop1 apps]$ ln -s kafka_2.11-0.8.2.0/ kafka

6．修改环境变量

[hadoop@hadoop1 ~]$ vi .bashrc

```
#Kafka
export KAFKA_HOME=/home/hadoop/apps/kafka
export PATH=$PATH:$KAFKA_HOME/bin
```

保存设置使其立即生效：

[hadoop@hadoop1 ~]$ source ~/.bashrc

6.1.4　Kafka 服务的启动

1．启动服务

（1）首先启动 ZooKeeper 集群，所有 ZooKeeper 节点都需要执行

[hadoop@hadoop1 ~]$ zkServer.sh start

（2）启动 Kafka 集群服务

[hadoop@hadoop1 kafka]$ bin/kafka-server-start.sh config/server.properties

2．创建消息

（1）创建的 Topic

[hadoop@hadoop1 kafka]$ bin/kafka-topics.sh --create --zookeeper hadoop1:2181 --replication-factor 3 --partitions 3 --topic mytopic

（2）查看 Topic 副本信息

[hadoop@hadoop1 kafka]$ bin/kafka-topics.sh --describe --zookeeper hadoop1:2181 --topic mytopic

（3）查看已经创建的 Topic 信息

[hadoop@hadoop1 kafka]$ bin/kafka-topics.sh --list --zookeeper hadoop1:2181

（4）生产者发送消息

[hadoop@hadoop1 kafka]$ bin/kafka-console-producer.sh --broker-list hadoop1:9092 --topic mytopic

（5）消费者消费消息

[hadoop@hadoop2 kafka]$ bin/kafka-console-consumer.sh --zookeeper hadoop1:2181 --from-beginning --topic mytopic

6.2 Kafka 组件术语

6.2.1 主题与日志

1. 主题

主题是存储消息的一个逻辑概念，可以简单理解为一类消息的集合，由使用方去创建。Kafka 中一般会有多个消费者去消费对应主题的消息，也可以存在多个生产者向主题中写入消息，如图 6-4 所示。

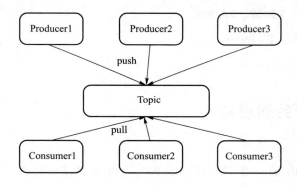

图 6-4 Kafka Topic

每个主题又可以划分成多个分区，每个分区存储不同的消息。当消息添加至分区时，会为其分配一个位移 offset（从 0 开始递增），并保证分区上唯一，消息在分区上的顺序由 offset 保证，即同一个分区内的消息是有序的，如图 6-5 所示。

同一个主题的不同分区会分配在不同的节点上，分区时保证 Kafka 集群具有水平扩展的基础。

2. 日志

以主题 nginx_access_log 为例，分区数为 3，如图 6-6 所示。分区在逻辑上对应一个日志（Log），在物理上对应的是一个文件夹。

```
drwxr-xr-x  2 root root 4096 10月 11 20:07 nginx_access_log-0/
drwxr-xr-x  2 root root 4096 10月 11 20:07 nginx_access_log-1/
drwxr-xr-x  2 root root 4096 10月 11 20:07 nginx_access_log-2/
```

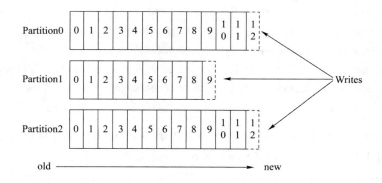

图 6-5　Topic 位移 offset

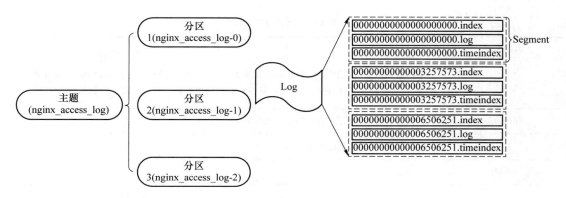

图 6-6　Topic 分区

消息写入分区时,实际上是将消息写入分区所在的文件夹中。日志又分成多个分片(Segment),每个分片由日志文件与索引文件组成,每个分片大小是有限的(在 Kafka 集群的配置文件 log.segment.bytes 中配置,默认为 1 073 741 824 B,即 1 GB),当分片大小超过限制则会重新创建一个新的分片,外界消息只会写入最新的一个分片(顺序 I/O)。日志文件分片类似于如下结构:

```
-rw-r--r-- 1 root root     1835920 10月 11 19:18 00000000000000000000.index
-rw-r--r-- 1 root root  1073741684 10月 11 19:18 00000000000000000000.log
-rw-r--r-- 1 root root     2737884 10月 11 19:18 00000000000000000000.timeindex
-rw-r--r-- 1 root root     1828296 10月 11 19:30 00000000000003257573.index
-rw-r--r-- 1 root root  1073741513 10月 11 19:30 00000000000003257573.log
-rw-r--r-- 1 root root     2725512 10月 11 19:30 00000000000003257573.timeindex
-rw-r--r-- 1 root root     1834744 10月 11 19:42 00000000000006506251.index
-rw-r--r-- 1 root root  1073741771 10月 11 19:42 00000000000006506251.log
-rw-r--r-- 1 root root     2736072 10月 11 19:42 00000000000006506251.timeindex
-rw-r--r-- 1 root root     1832152 10月 11 19:54 00000000000009751854.index
-rw-r--r-- 1 root root  1073740984 10月 11 19:54 00000000000009751854.log
-rw-r--r-- 1 root root     2731572 10月 11 19:54 00000000000009751854.timeindex
-rw-r--r-- 1 root root     1808792 10月 11 20:06 00000000000012999310.index
-rw-r--r-- 1 root root  1073741584 10月 11 20:06 00000000000012999310.log
-rw-r--r-- 1 root root          10 10月 11 19:54 00000000000012999310.snapshot
```

```
-rw-r--r--  1 root root    2694564 10月 11 20:06 00000000000012999310.timeindex
-rw-r--r--  1 root root   10485760 10月 11 20:09 00000000000016260431.index
-rw-r--r--  1 root root  278255892 10月 11 20:09 00000000000016260431.log
-rw-r--r--  1 root root         10 10月 11 20:06 00000000000016260431.snapshot
-rw-r--r--  1 root root   10485756 10月 11 20:09 00000000000016260431.timeindex
-rw-r--r--  1 root root          8 10月 11 19:03 leader-epoch-checkpoint
```

一个分片包含多个后缀不同的日志文件,分片中第一个消息的 offset 将作为该分片的基准偏移量,偏移量固定长度为 20 位,不够则前面补齐 0,然后将其作为索引文件以及日志文件的文件名,如 00000000000003257573.index、00000000000003257573.log、00000000000003257573.timeindex,文件名相同的文件组成一个分片(忽略后缀名)。日志文件说明如表 6-1 所示。

表 6-1　日志文件说明

文件类型	作用
.index	偏移量索引文件,记录<相对位移,起始地址>映射关系,其中相对位移表示该分片的消息的偏移量,从 0 开始计算,起始地址表示对应相对位移消息在分片 .log 文件中的起始地址
.timeindex	时间戳索引文件,记录<时间戳,相对位移>映射关系
.log	日志文件,存储消息的详细信息
.snaphot	快照文件
.deleted	分片文件删除时会先将该分片的所有文件加上 .deleted 后缀,然后由 delete-file 任务延迟删除这些文件(file.delete.delay.ms 可以设置延迟删除的时间)
.cleaned	日志清理时的临时文件
.swap	Log Compaction 之后的临时文件
.leader-epoch-checkpoint	在日志截断之前,加一层判断是否有必要截断。而这个判断的依据就是 Leader Epoch

3. 日志索引

首先介绍 .index 文件,这里以文件 00000000000003257573.index 为例,首先我们可以查看该索引文件的内容,可以看到输出结构为< offset, position >,实际上索引文件中保存的并不是 offset,而是相对位移,例如,第一条消息的相对位移为 0,格式化输出时加上了基准偏移量。

如图 6-7 所示,< 114,17413 >表示该分片相对位移为 114 的消息,其位移为 3257573＋114,即 3257687,position 表示对应 offset 在 .log 文件中的物理地址,即通过 .index 索引文件可以获取对应 offset 所在的物理地址。

索引采用稀疏索引的方式构建,并不保证分片中的每条消息都在索引文件中有映射关系(.timeindex 索引类似),主要是为了节省磁盘空间、内存空间,因为索引文件最终会映射到内存中。

我们可以通过相关命令来查看该分片索引文件的前 10 条记录,如下所示:

```
$ bin/kafka-dump-log.sh --files /tmp/kafka-logs/nginx_access_log-1/00000000000003257573.index | head -n 10
```

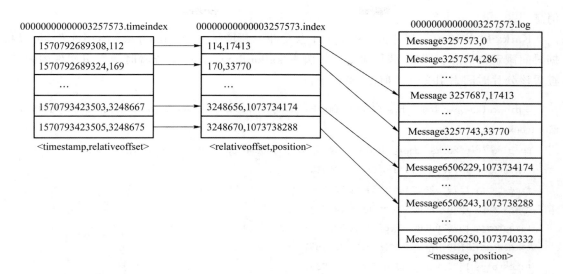

图 6-7　日志索引

```
Dumping /tmp/kafka-logs/nginx_access_log-1/00000000000003257573.index
offset: 3257687 position: 17413
offset: 3257743 position: 33770
offset: 3257799 position: 50127
offset: 3257818 position: 66484
offset: 3257819 position: 72074
offset: 3257871 position: 87281
offset: 3257884 position: 91444
offset: 3257896 position: 95884
offset: 3257917 position: 100845
```

查看该分片索引文件的后 10 条记录：

$ bin/kafka-dump-log.sh --files /tmp/kafka-logs/nginx_access_log-1/00000000000003257573.index | tail -n 10

```
offset: 6506124 position: 1073698512
offset: 6506137 position: 1073702918
offset: 6506150 position: 1073707263
offset: 6506162 position: 1073711499
offset: 6506176 position: 1073716197
offset: 6506188 position: 1073720433
offset: 6506205 position: 1073725654
offset: 6506217 position: 1073730060
offset: 6506229 position: 1073734174
offset: 6506243 position: 1073738288
```

示例：查看 offset 为 6506155 的消息。首先根据 offset 找到对应的分片，6506155 所对应的分片为 00000000000003257573，然后通过二分法在 00000000000003257573.index 文件中找到不大于 6506155 的最大索引值，得到 < offset：6506150，position：1073707263 >，最后从 00000000000003257573.log 的 1073707263 位置开始顺序扫描，找到 offset 为 6506155 的

消息。

　　Kafka 从 0.10.0.0 版本起，为分片日志文件中新增了一个 .timeindex 索引文件，可以根据时间戳定位消息。同样，我们可以通过脚本 kafka-dump-log.sh 查看时间索引文件的内容。查看该分片时间索引文件的前 10 条记录：

```
$ bin/kafka-dump-log.sh --files /tmp/kafka-logs/nginx_access_log-1/00000000000003257573.timeindex | head -n 10
```

```
Dumping /tmp/kafka-logs/nginx_access_log-1/00000000000003257573.timeindex
timestamp: 1570792689308 offset: 3257685
timestamp: 1570792689324 offset: 3257742
timestamp: 1570792689345 offset: 3257795
timestamp: 1570792689348 offset: 3257813
timestamp: 1570792689357 offset: 3257867
timestamp: 1570792689361 offset: 3257881
timestamp: 1570792689364 offset: 3257896
timestamp: 1570792689368 offset: 3257915
timestamp: 1570792689369 offset: 3257927
```

　　查看该分片时间索引文件的后 10 条记录：

```
$ bin/kafka-dump-log.sh --files /tmp/kafka-logs/nginx_access_log-1/00000000000003257573.timeindex | tail -n 10
```

```
timestamp: 1570793423474 offset: 6506136
timestamp: 1570793423477 offset: 6506150
timestamp: 1570793423481 offset: 6506159
timestamp: 1570793423485 offset: 6506176
timestamp: 1570793423489 offset: 6506188
timestamp: 1570793423493 offset: 6506204
timestamp: 1570793423496 offset: 6506214
timestamp: 1570793423500 offset: 6506228
timestamp: 1570793423503 offset: 6506240
timestamp: 1570793423505 offset: 6506248
```

　　示例：查看时间戳 1570793423501 开始的消息。

　　① 首先定位分片，将 1570793423501 与每个分片的最大时间戳进行对比（最大时间戳取时间索引文件的最后一条记录时间，如果时间为 0 则取该日志分片的最近修改时间），直到找到大于或等于 1570793423501 的日志分片，因此会定位到时间索引文件 00000000000003257573.timeindex，其最大时间戳为 1570793423505。

　　② 通过二分法找到大于或等于 1570793423501 的最大索引项，即 < timestamp：1570793423503 offset：6506240 >（6506240 为 offset，相对位移为 3248667）。

　　③ 根据相对位移 3248667 去索引文件中找到不大于该相对位移的最大索引值，即 < 3248656,1073734174 >。

　　④ 从日志文件 00000000000003257573.log 的 1073734174 位置处开始扫描，查找不小于 1570793423501 的数据。

6.2.2 Kafka 日志处理

与其他消息中间件有所不同的是,Kafka 集群中的消息不会因为消费与否而删除,和系统日志一样,Kafka 消息最终会存储在硬盘中,并提供对应的策略周期性(通过参数 log.retention.check.interval.ms 来设置,默认为 5 min)执行删除或者压缩操作(Broker 配置文件 log.cleanup.policy 参数如果为"delete"则执行删除操作,如果为"compact"则执行压缩操作,默认为"delete")。

1. 基于时间的日志删除

基于时间的日志删除如表 6-2 所示。

表 6-2 基于时间的日志删除

参数	默认值	说明
log.retention.hours	168	日志保留时间(小时)
log.retention.minutes	无	日志保留时间(分钟),优先级大于小时
log.retention.ms	无	日志保留时间(毫秒),优先级大于分钟

若消息在集群中的保留时间超过设定阈值(log.retention.hours,默认为 168 h,即 7 天),则需要进行删除。这里会根据分片日志的最大时间戳来判断该分片的时间是否满足删除条件,最大时间戳会选取时间索引文件中的最后一条索引记录,如果对应的时间戳值大于 0 则取该值,否则为最近一次修改时间,如图 6-8 所示。

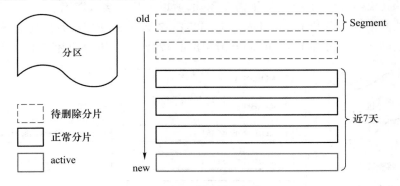

图 6-8 保留的时间阈值

这里不直接选取最后修改时间的原因是避免分片日志的文件被无意修改而导致其时间不准。如果恰好该分区下的所有日志分片均已过期,那么会先生成一个新的日志分片作为新消息的写入文件,然后再执行删除操作。

2. 基于空间的日志删除

基于空间的日志删除如表 6-3 所示。

表 6-3 基于空间的日志删除

参数	默认值	说明
log.retention.bytes	1 073 741 824(即 1 GB),默认未开启,即无穷大	日志文件总大小,并不是指单个分片的大小
log.segment.bytes	1 073 741 824(即 1 GB)	单个日志分片大小

首先会计算待删除的日志大小 diff(totalSize－log.retention.bytes)，然后从最旧的一个分片开始查看可以执行删除操作的文件集合（如果 diff－segment.size≥0，则满足删除条件），最后执行删除操作，如图 6-9 所示。

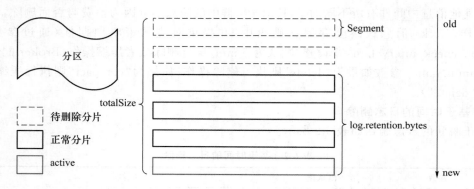

图 6-9 基于空间的阈值

3. 基于日志起始偏移量的日志删除

一般情况下，日志文件的起始偏移量（logStartOffset）会等于第一个日志分片的 baseOffset，但是其值会因为删除消息请求而增长，logStartOffset 的值实际上是日志集合中的最小消息，而小于这个值的消息都会被清理掉。如图 6-10 所示，我们假设 logStartOffset＝7421048，日志删除流程如下。

- 从最旧的日志分片开始遍历，判断其下一个分片的 baseOffset 是否小于或等于 logStartOffset 值，如果满足，则需要删除，所以分片一会被删除。
- 分片二的下一个分片 baseOffset＝6506251＜7421048，所以分片二也会被删除。
- 分片三的下一个分片 baseOffset＝9751854＞7421048，所以分片三不会被删除。

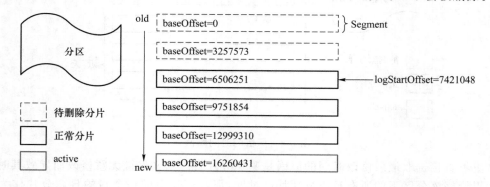

图 6-10 基于日志起始偏移量的日志删除

4. 日志压缩

前面提到当 Broker 配置文件 log.cleanup.policy 参数值设置为"compact"时，会执行压缩操作，这里的压缩和普通意义上的压缩不一样，这里的压缩是指对相同 key 的消息只保留最后一个版本的 value，如图 6-11 所示，压缩之前 offset 是连续递增的，压缩之后 offset 递增可能不连续，只保留 5 条消息记录。

Kafka 日志目录下的 cleaner-offset-checkpoint 文件用来记录每个主题的每个分区中已经清理的偏移量，通过这个偏移量可以将分区中的日志文件分成两个部分：clean 表示已经压缩

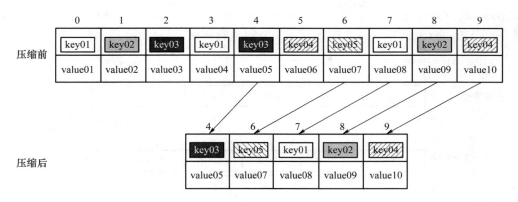

图 6-11 日志压缩

过；dirty 表示还未进行压缩。如图 6-12 所示（active segment 不会参与日志的压缩操作，因为会有新的数据写入该文件）。

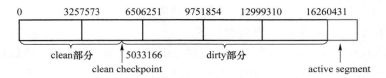

图 6-12 cleaner-offset-checkpoint 文件

例如，我们有时会看到如下日志及 checkpoint 文件内容：

```
-rw-r--r--  1 root root     4 10月 11 19:02 cleaner-offset-checkpoint
drwxr-xr-x  2 root root  4096 10月 11 20:07 nginx_access_log-0/
drwxr-xr-x  2 root root  4096 10月 11 20:07 nginx_access_log-1/
drwxr-xr-x  2 root root  4096 10月 11 20:07 nginx_access_log-2/
-rw-r--r--  1 root root     0  9月 18 09:50 .lock
-rw-r--r--  1 root root     4 10月 16 11:19 log-start-offset-checkpoint
-rw-r--r--  1 root root    54  9月 18 09:50 meta.properties
-rw-r--r--  1 root root  1518 10月 16 11:19 recovery-point-offset-checkpoint
-rw-r--r--  1 root root  1518 10月 16 11:19 replication-offset-checkpoint
# cat cleaner-offset-checkpoint
nginx_access_log 0 5033168
nginx_access_log 1 5033166
nginx_access_log 2 5033168
```

日志压缩时会根据 dirty 部分数据占日志文件的比例（cleanableRatio）来判断优先压缩的日志，然后为 dirty 部分的数据建立 key 与 offset 映射关系（保存对应 key 的最大 offset），存入 SkimpyoffsetMap 中，然后复制 Segment 中的数据，只保留 SkimpyoffsetMap 中记录的消息。

压缩之后的相关日志文件会变小，为了避免出现过小的日志文件与索引文件，压缩时会对所有的 Segment 进行分组（一个组的分片大小不会超过设置的 log.segment.bytes 值大小），同一个分组的多个日志分片压缩之后变成一个分片。

如图 6-13 所示，所有消息都还没压缩前 clean checkpoint 值为 0，表示该分区的数据还没进行压缩。第一次压缩后，之前每个分片的日志文件大小都有所减少，同时会移动 clean checkpoint 的位置到这一次压缩结束的 offset 值。第二次压缩时，会将分片{0.5 GB,0.4 GB}组成一个分组，将{0.7 GB,0.2 GB}组成一个分组进行压缩，以此类推。

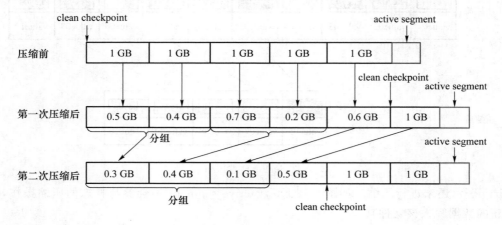

图 6-13 日志文件压缩前后对比

如图 6-14 所示，日志压缩的主要流程如下。

① 计算 deleteHorizonMs 值：当某个消息的 value 为空时，该消息会被保留一段时间，超时之后会在下一次的日志压缩中被删除，所以这里会计算 deleteHorizonMs，根据该值确定可以删除 value 为空的日志分片（deleteHorizonMs = clean 部分的最后一个分片的 lastModifiedTime－deleteRetentionMs，deleteRetentionMs 通过配置文件 log. cleaner. delete. retention. ms 配置，默认为 24 h）。

② 确定压缩 dirty 部分的 offset 范围[firstDirtyOffset,endOffset]：其中 firstDirtyOffset 表示 dirty 的起始位移，一般会等于 clean checkpoint 值，firstUncleanableOffset 表示不能清理的最小位移，一般会等于活跃分片的 baseOffset，然后从 firstDirtyOffset 位置开始遍历日志分片，并填充 key 与 offset 的映射关系至 SkimpyoffsetMap 中，当该 Map 被填充满或到达上限 firstUncleanableOffset 时，就可以确定日志压缩上限 endOffset。

③ 将（logStartOffset,endOffset）中的日志分片进行分组，然后按照分组的方式进行压缩。

图 6-14 日志压缩流程

6.2.3 消息副本

Kafka 支持消息的冗余备份，可以设置对应主题的副本数（--replication-factor 参数设置主题的副本数，可在创建主题的时候指定，offsets. topic. replication. factor 参数设置消费主题

_consumer_offsets 副本数，默认为 3），每个副本包含的消息一样（但不是完全一致，从副本的数据和主副本的数据相比，可能稍微有些落后）。

每个分区的副本集合中会有一个副本被选举为主副本（Leader），其他为从副本，所有的读写请求由主副本对外提供，从副本负责将主副本的数据同步到自己所属分区，如果主副本所在分区宕机，则会重新选举出新的主副本对外提供服务。

1. ISR 集合

ISR 集合表示目前可用的副本集合，每个分区中的 Leader 副本会维护此分区的 ISR 集合。这里的可用是指从副本的消息量与主副本的消息量相差不大，加入 ISR 集合中的副本必须满足以下几个条件。

① 副本所在节点需要与 ZooKeeper 维持心跳。

② 从副本的最后一条消息的 offset 与主副本的最后一条消息的 offset 差值不超过设定阈值（replica.lag.max.messages），或者从副本的 LEO 落后于主副本的 LEO 时长不大于设定阈值（replica.lag.time.max.ms），官方推荐使用后者判断，并在 Kafka 0.10.0.0 版本中移除了 replica.lag.max.messages 参数。

如果从副本不满足以上的任意条件，则会将其踢出 ISR 集合，当其再次满足以上条件之后又会被重新加入集合中。ISR 的引入主要是解决同步副本与异步副本两种方案各自的缺陷（同步副本中如果有一个副本宕机或者超时就会拖慢该副本组的整体性能；如果仅使用异步副本，当所有的副本消息均远落后于主副本时，一旦主副本宕机而重新选举，就会存在消息丢失情况）。

2. HW 和 LEO

HW（High Watermark）是一个比较特殊的 offset 标记，消费端消费时只能拉取到小于 HW 的消息，而 HW 及之后的消息对于消费者是不可见的，该值由主副本管理，当 ISR 集合中的全部从副本都拉取到 HW 指定消息之后，主副本会将 HW 值加 1，即指向下一个 offset 位移，这样可以保证 HW 之前消息的可靠性，如图 6-15 所示。

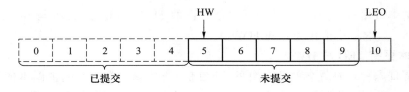

图 6-15　High Watermark

LEO（Log End Offset）（见图 6-16）表示当前副本最新消息的下一个 offset，所有副本都存在这样一个标记，如果是主副本，当生产端向其追加消息时，会将 LEO 值加 1。当从副本从主副本成功拉取到消息时，其值也会增加。

3. 从副本更新 LEO 与 HW

从副本更新 LEO 与 HW 如图 6-17 所示。

从副本的数据来自主副本，从副本通过向主副本发送 fetch 请求获取数据，从副本的 LEO 值会保存在两个地方，一个是自身所在节点，一个是主副本所在节点，自身所在节点保存 LEO 主要是为了更新自身的 HW 值，主副本保存从副本的 LEO 也是为了更新其 HW 值。

从副本每写入一条新消息就会增加自身的 LEO 值，主副本收到从副本的 fetch 请求，会先从自身的日志中读取对应数据，在数据返回给从副本之前会先更新其保存的从副本 LEO

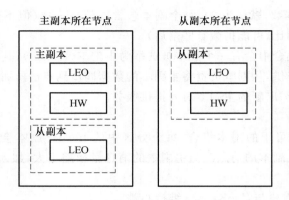

图 6-16 Log End Offset

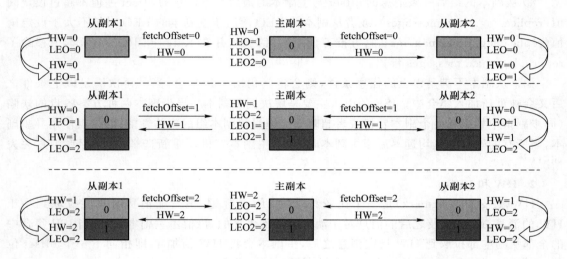

图 6-17 从副本更新 LEO 与 HW

值。一旦从副本数据写入完成,就会尝试更新自己的 HW 值,比较 LEO 与 fetch 响应中主副本返回的 HW 值,取最小值作为新的 HW 值。

4. 主副本更新 LEO 与 HW

主副本有日志写入时就会更新自身的 LEO 值,与从副本类似。而主副本的 HW 值是分区的 HW 值,决定分区数据对于消费端的可见性,在以下 4 种情况下,主副本会尝试更新其 HW 值。

① 副本成为主副本:当某个副本成为主副本时,Kafka 会尝试更新分区的 HW 值。

② Broker 出现崩溃导致副本被踢出 ISR 集合:如果有 Broker 节点崩溃则要看是否影响对应分区,然后会去检查分区的 HW 值是否需要更新。

③ 生成端向主副本写入消息时:消息写入会增加其 LEO 值,此时会查看是否需要修改 HW 值。

④ 主副本接收到从副本的 fetch 请求时:主副本在处理从副本的 fetch 请求时会尝试更新分区的 HW 值。

以上是去尝试更新 HW,但是不一定会更新,主副本上保存着从副本的 LEO 值与自身的 LEO 值,这里会比较所有满足条件的副本的 LEO 值,并选择最小的 LEO 值作为分区的 HW

值,其中满足条件的副本是指满足以下两个条件之一:

① 副本在 ISR 集合中。

② 副本的 LEO 落后于主副本的 LEO 时长不大于设定阈值(replica.lag.time.max.ms,默认为 10 s)。

6.2.4 数据处理场景

1. 数据丢失场景

前面提到,如果只依赖 HW 来进行日志截断以及水位的判断会存在问题。如图 6-18 所示,假定存在两个副本 A、B,最开始 A 为主副本,B 为从副本,且参数 min.insync.replicas=1,即 ISR 只有一个副本时也会返回成功:

① 初始情况为主副本 A 已经写入了两条消息,对应 HW=1,LEO=2,LEOB=1,从副本 B 写入了一条消息,对应 HW=1,LEO=1。

② 此时从副本 B 向主副本 A 发起 fetchOffset=1 请求,主副本收到请求之后更新 LEOB=1,表示副本 B 已经收到了消息 0,然后尝试更新 HW 值,min(LEO,LEOB)=1,即不需要更新,然后将消息 1 以及当前分区 HW=1 返回给从副本 B,从副本 B 收到响应之后写入日志并更新 LEO=2,然后更新其 HW=1,虽然已经写入了两条消息,但是其 HW 值在下一轮的请求中才会更新为 2。

③ 此时从副本 B 重启,重启之后会根据 HW 值进行日志截断,即消息 1 会被删除。

④ 从副本 B 向主副本 A 发送 fetchOffset=1 请求,如果此时主副本 A 没有什么异常,则和第②步一样,没有什么问题,假设此时主副本宕机了,那么从副本 B 会变成主副本。

⑤ 副本 A 恢复之后会变成从副本,并根据 HW 值进行日志截断,即把消息 1 丢失,此时消息 1 就永久丢失了。

2. 数据不一致场景

如图 6-19 所示,假定存在两个副本 A、B,最开始 A 为主副本,B 为从副本,且参数 min.insync.replicas=1,即 ISR 只有一个副本时也会返回成功:

① 初始状态为主副本 A 已经写入了两条消息,对应 HW=1,LEO=2,LEOB=1,从副本 B 也同步了两条消息,对应 HW=1,LEO=2。

② 此时从副本 B 向主副本 A 发送 fetchOffset=2 请求,主副本 A 在收到请求后更新分区 HW=2 并将该值返回给从副本 B,如果此时从副本 B 宕机则会导致 HW 值写入失败。

③ 假设此时主副本 A 也宕机了,从副本 B 先恢复并成为主副本,此时会发生日志截断,只保留消息 0,然后对外提供服务,假设外部写入了一个消息 1(这个消息与之前的消息 1 不一样,用不同的颜色标识不同消息)。

④ 副本 A 恢复之后会变成从副本,不会发生日志截断,因为 HW=2,但是对应位移 1 的消息其实是不一致的。

3. leader epoch 机制

HW 值被用作衡量副本备份成功与否以及出现失败情况时的日志截断依据可能会导致数据丢失与数据不一致情况,因此在 Kafka 0.11.0.0 中引入了 leader epoch 概念。

leader epoch 表示一个键值对< epoch,offset >,其中 epoch 表示 Leader 主副本的版本号,从 0 开始编码,Leader 每变更一次就会加 1,offset 表示该 epoch 版本的主副本写入第一条消息的位置。

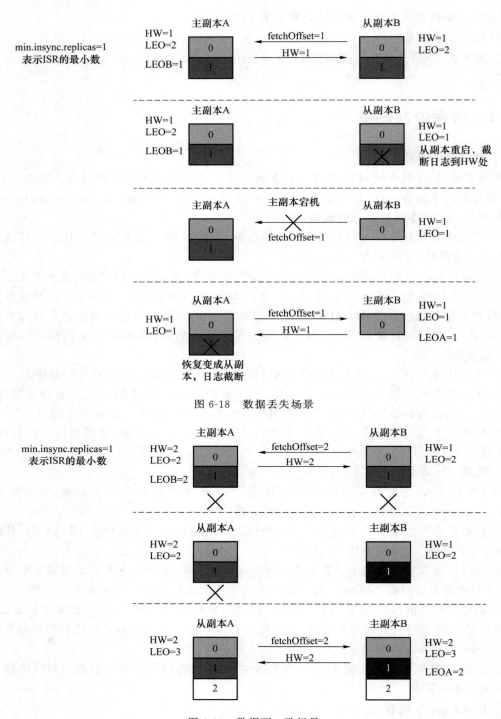

图 6-18　数据丢失场景

图 6-19　数据不一致场景

例如,<0,0>表示第一个主副本从位移 0 开始写入消息,<1,100>表示第二个主副本版本号为 1 并从位移 100 开始写入消息,主副本会将该信息保存在缓存中并定期写入 checkpoint 文件中,每次发生主副本切换时都会从缓存中查询该信息,下面简单介绍 leader epoch 的工作原理。

① 每条消息都会包含一个 4 字节的 leader epoch 值。

② 每个 log 目录都会创建一个 leader epoch sequence 文件,用来存放主副本版本号以及开始位移。

③ 当一个副本成为主副本之后,会在 leader epoch sequence 文件末尾添加一条新的记录,然后每条新的消息就会变成新的 leader epoch 值。

④ 当某个副本宕机重启之后,会进行以下操作:

- 从 leader epoch sequence 文件中恢复所有的 leader epoch。
- 向分区主副本发送 LeaderEpoch 请求,请求包含了从副本的 leader epoch sequence 文件中的最新 leader epoch 值。
- 主副本返回从副本对应 leader epoch 的 lastOffset,返回的 lastOffset 分为两种情况,一种是返回比从副本请求中的 leader epoch 版本大 1 的开始位移,另外一种是与请求中的 leader epoch 相等则直接返回当前主副本的 LEO 值。
- 如果从副本的 leader epoch 开始位移大于从主副本中返回的 lastOffset,那么会将从副本的 leader epoch 值保持和主副本一致。
- 从副本截断本地消息到主副本返回的 lastOffset 所在位移处。
- 从副本开始从主副本中拉取数据。
- 在获取数据时,如果从副本发现消息中的 leader epoch 值比自身的最新 leader epoch 值大,则会将该 leader epoch 值写入 leader epoch sequence 文件,然后继续同步文件。

下面介绍 leader epoch 机制如何避免前面提到的两种异常场景。

4. 数据丢失场景解决

① 如图 6-20 所示,从副本 B 重启之后向主副本 A 发送 offsetsForLeaderEpochRequest,epoch 主、从副本相等,则主副本 A 返回当前的 LEO=2,从副本 B 中没有任何大于 2 的位移,因此不需要截断。

② 当从副本 B 向主副本 A 发送 fetchOffset=2 请求时,主副本 A 宕机,所以从副本 B 成为主副本,并更新 leader epoch 值为< epoch=1, offset=2 >,HW 值更新为 2。

③ A 恢复之后成为从副本,并向 B 发送 fetchOffset=2 请求,B 返回 HW=2,则从副本 A 更新 HW=2。

④ 主副本 B 接收外界的写请求,从副本 A 向主副本 B 不断发起数据同步请求。

从图 6-20 中可以看出,引入 leader epoch 值之后避免了前面提到的数据丢失情况,但是这里需要注意的是,如果在上面的第①步中,从副本 B 重启之后向主副本 A 发送 offsetsForLeaderEpochRequest 请求失败,即主副本 A 宕机了,那么消息 1 就会丢失,具体可见下面的数据不一致场景。

5. 数据不一致场景解决

① 如图 6-21 所示,从副本 B 恢复之后向主副本 A 发送 offsetsForLeaderEpochRequest 请求,由于主副本宕机了,因此副本 B 将变成主副本并将消息 1 截断,此时接收到新消息 1 的写入。

② 副本 A 恢复之后变成从副本并向主副本 B 发送 offsetsForLeaderEpochRequest 请求,请求的 epoch 值小于主副本 B 的,因此主副本 B 会返回 epoch=1 时的开始位移,即 lastOffset=1,因此从副本 A 会截断消息 1。

③ 从副本 A 从主副本 B 拉取消息,并更新 leader epoch 值< epoch=1, offset=1 >。

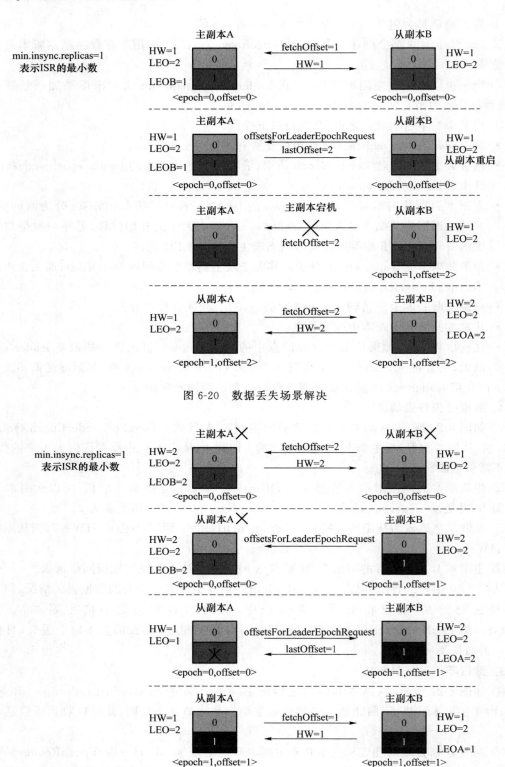

图 6-20　数据丢失场景解决

图 6-21　数据不一致场景解决

可以看出，leader epoch 的引入避免了数据不一致，但是若两个副本均宕机，则还是存在数据丢失的场景。前面的所有讨论都建立在 min.insync.replicas=1 的前提下，因此需要在数据的可靠性与速度方面做权衡。

6.2.5 生产者

1. 消息分区选择

生产者的作用主要是生产消息，将消息存入 Kafka 对应主题的分区中，具体某个消息应该存入哪个分区，由以下 3 个策略决定（优先级由上到下依次递减）。

① 如果消息发送时指定了消息所属分区，则会直接发往指定分区。

② 如果没有指定消息分区，但是设置了消息的 key，则会根据 key 的哈希值选择分区。

③ 如果前两者均不满足，则会采用轮询的方式选择分区。

2. ack 参数的设置及意义

生产端向 Kafka 集群发送消息时，可以通过 request.required.acks 参数来设置数据的可靠性级别。

① 1：默认为 1，表示在 ISR 中的 Leader 副本成功接收到数据并确认后再发送下一条消息，如果主节点宕机则可能出现数据丢失场景，详细分析可参考前述内容。

② 0：表示生产端不需要等待节点的确认就可以继续发送下一批数据，这种情况下数据传输效率最高，但是数据的可靠性最低。

③ -1：表示生产端需要等待 ISR 中的所有副本节点都收到数据之后才算消息写入成功，可靠性最高，但是性能最低，如果服务端的 min.insync.replicas 值设置为 1，那么在这种情况下允许 ISR 集合只有一个副本，因此也会存在数据丢失的情况。

3. 幂等特性

所谓的幂等性，是指一次或者多次请求某一个资源对于资源本身应该具有同样的结果（网络超时等问题除外），通俗一点的理解就是同一个操作任意执行多次产生的影响或效果与执行一次产生的影响或效果相同，幂等的关键在于服务端能否识别出请求是否重复，然后过滤这些重复请求，通常情况下需要以下信息来实现幂等特性。

① 唯一标识：判断某个请求是否重复需要有一个唯一标识，然后服务端就能根据这个唯一标识来判断是否为重复请求。

② 记录已经处理过的请求：服务端需要记录已经处理过的请求，然后根据唯一标识来判断是否为重复请求，如果已经处理过，则直接拒绝或者不做任何操作返回成功。

Kafka 中 Producer 端的幂等性是指当发送同一条消息时，消息在集群中只会被持久化一次，其幂等在以下条件中才成立。

① 只能保证生产端在单个会话内的幂等，如果生产端因为某些原因意外挂掉然后重启，此时是没办法保证幂等的，因为这时没办法获取到之前的状态信息，即无法做到跨会话级别的幂等。

② 幂等性不能跨多个主题分区，只能保证单个分区内的幂等，涉及多个消息分区时，中间的状态并没有同步。

如果要支持跨会话或者跨多个消息分区的情况，则需要使用 Kafka 的事务性来实现。

为了实现生产端的幂等语义，引入了 Producer ID 与 Sequence Number 的概念，如下。

① Producer ID(PID)：每个生产者在初始化时都会分配一个唯一的 PID，PID 的分配对于用户来说是透明的。

② Sequence Number(序列号)：对于给定的 PID 而言，序列号从 0 开始单调递增，每个主题分区均会产生一个独立序列号，生产者在发送消息时会给每条消息添加一个序列号。Broker 端缓存了已经提交的消息的序列号，只有比缓存分区中最后提交的消息的序列号大 1 的消息才会被接收，其他会被拒绝。

4．生产端消息发送流程

下面简单介绍支持幂等的消息发送端的工作流程。

① 生产端通过 KafkaProducer 会将数据添加到 RecordAccumulator 中，添加数据时会判断是否需要新建一个 ProducerBatch(以下简写为 Batch)。

② 生产端后台启动发送线程，会判断当前的 PID 是否需要重置，重置的原因是某些消息分区的 Batch 重试多次仍然失败，最后因为超时而被移除，这个时候序列号无法连续，导致后续消息无法发送，因此会重置 PID，并将相关缓存信息清空，这个时候消息会丢失。

③ 发送线程判断是否需要新申请 PID，如果需要则会阻塞直到获取到 PID 信息。

④ 发送线程在调用 sendProducerData() 方法发送数据时，会进行以下判断：

- 判断主题分区是否可以继续发送、PID 是否有效，如果是重试 Batch 需要判断之前的 Batch 是否发送完成，如果没有发送完成则会跳过当前主题分区的消息发送，直到前面的 Batch 发送完成。
- 如果对应 Batch 没有分配对应的 PID 与序列号信息，则会在这里进行设置。

5．服务端消息接收流程

服务端(Broker)在收到生产端发送的数据写请求之后，会进行一些判断来决定是否可以写入数据，这里主要介绍幂等相关的操作流程。

① 如果请求设置了幂等特性，则会检查是否对 ClusterResource 有 IdempotentWrite 权限，如果没有，则会返回错误 CLUSTER_AUTHORIZATION_FAILED。

② 检查是否有 PID 信息。

③ 根据 Batch 的序列号检查该 Batch 是否重复，服务端会缓存每个 PID 对应主题分区的最近 5 个 Batch 信息，如果有重复，则直接返回写入成功，但是不会执行真正的数据写入操作。

④ 如果有 PID 且 Batch 非重复，则进行以下操作：

- 判断该 PID 是否已经存在缓存中。
- 如果不存在则判断序列号是否从 0 开始，如果是则表示为新的 PID，在缓存中记录 PID 的信息(包括 PID、epoch 以及序列号信息)，然后执行数据写入操作；如果不存在但是序列号不是从 0 开始，则直接返回错误，表示 PID 在服务端已经过期或者 PID 写的数据已经过期。
- 如果 PID 存在，则会检查 PID 的 epoch 版本是否与服务端一致，如果不一致且序列号不是从 0 开始，则返回错误，如果 epoch 不一致但是序列号是从 0 开始，则可以正常写入。
- 如果 epoch 版本一致，则会查询缓存中最近一次序列号是否连续，不连续则会返回错误，否则正常写入。

6.2.6 消费者

消费者主要是从 Kafka 集群拉取消息,然后进行相关的消费逻辑,消费者的消费进度由其自身控制,增加消费的灵活性,例如,消费端可以控制重复消费某些消息或者跳过某些消息进行消费。

1. 消费组

多个消费者可以组成一个消费组,每个消费者只属于一个消费组。消费组订阅主题的每个分区只会分配给该消费组中的某个消费者处理,不同的消费组之间彼此隔离无依赖。同一条消息只会被消费组中的一个消费者消费,如果想要让同一条消息被多个消费者消费,那么每个消费者需要属于不同的消费组,且对应消费组中只有一个消费者,消费组的引入可以实现消费的"独占"或"广播"效果。消费组特点总结如下。

① 消费组下可以有多个消费者,个数支持动态变化。
② 消费组订阅主题下的每个分区只会分配给消费组中的一个消费者。
③ group.id 标识消费组,相同则属于同一消费组。
④ 不同消费组之间相互隔离、互不影响。

如图 6-22 所示,消费组 1 中包含两个消费者,其中消费者 1 分配消费分区 0,消费者 2 分配消费分区 1 与分区 2。此外,消费组的引入还支持消费者的水平扩展及故障转移,例如,从图 6-22 中我们可以看出消费者 2 的消费能力不足,相对于消费者 1 来说消费进度比较落后,我们可以往消费组里面增加一个消费者以提高整体的消费能力,如图 6-23 所示。

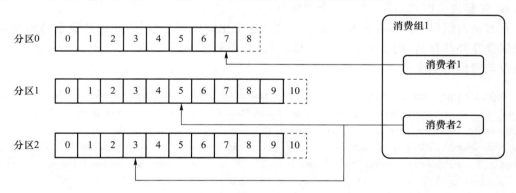

图 6-22 消费组

假设消费者 1 所在机器出现宕机,消费组会发送重平衡,假设将分区 0 分配给消费者 2 进行消费,如图 6-24 所示。同个消费组中消费者的个数不是越多越好,最大不能超过主题对应的分区数,如果超过则会出现超过的消费者分配不到分区的情况,因为分区一旦分配给消费者就不会再变动,除非组内消费者个数出现变动而发生重平衡。

2. 消费位移

(1) 消费位移主题

Kafka 0.9 开始将消费端的位移信息保存在集群的内部主题(__consumer_offsets)中,该主题默认为 50 个分区,每条日志项的格式都是< TopicPartition,OffsetAndMetadata >,其 key 为主题分区,主要存放主题、分区以及消费组信息,value 为 OffsetAndMetadata 对象,主要包

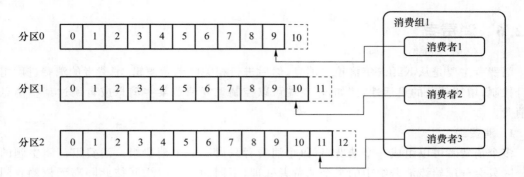

图 6-23　添加消费者

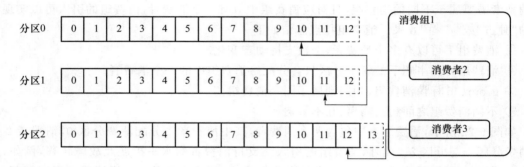

图 6-24　消费组重平衡

括位移、位移提交时间、自定义元数据等信息。

只有消费组向 Kafka 中提交位移时才会向这个主题中写入数据，如果消费端将消费位移信息保存在外部存储，则不会有消费位移信息，下面可以通过 kafka-console-consumer.sh 脚本查看消费位移主题信息：

```
# bin/kafka-console-consumer.sh --topic __consumer_offsets --bootstrap-server localhost:9092
--formatter "kafka.coordinator.group.GroupMetadataManager\$OffsetsMessageFormatter" --consumer.
config config/consumer.properties --from-beginning
    [consumer-group01,nginx_access_log,2]::OffsetAndMetadata(offset = 17104625, leaderEpoch =
Optional.[0], metadata = , commitTimestamp = 1573475863555, expireTimestamp = None)[consumer-group01,
nginx_access_log,1]::OffsetAndMetadata(offset = 17103024, leaderEpoch = Optional.[0], metadata = ,
commitTimestamp = 1573475863555, expireTimestamp = None)[consumer-group01,nginx_access_log,0]::
OffsetAndMetadata(offset = 17107771, leaderEpoch = Optional.[0], metadata = , commitTimestamp =
1573475863555, expireTimestamp = None)
```

（2）消费位移自动提交

消费端可以通过设置参数 enable.auto.commit 来控制是自动提交还是手动提交，如果值为 true 则表示自动提交，在消费端的后台会定时地提交消费位移信息，时间间隔由 auto.commit.interval.ms（默认为 5 s）设置。但是如果设置为自动提交会存在以下几个问题。

① 可能存在重复的位移数据提交到消费位移主题中，因为每隔 5 s 会往主题中写入一条消息，不管是否有新的消费记录，这样就会产生大量的同 key 消息，其实只需要一条，因此需要依赖前面提到的日志压缩策略来清理数据。

② 重复消费，假设位移提交的时间间隔为 5 s，那么在 5 s 内如果发生了重平衡，则所有的

消费者会从上一次提交的位移处开始消费,期间消费的数据则会再次被消费。

(3) 消费位移手动提交

手动提交需要将 enable.auto.commit 参数设置为 false,然后由业务消费端来控制消费进度,手动提交又分为以下 3 种类型。

① 同步手动提交位移:如果调用的是同步提交方法 commitSync(),则会将拉取的最新位移提交到 Kafka 集群,提交成功前会一直等待提交成功。

② 异步手动提交位移:如果调用的是异步提交方法 commitAsync(),则在调用该方法之后会立刻返回,不会阻塞,然后可以通过回调函数执行相关的异常处理逻辑。

③ 指定提交位移:指定提交位移也分为异步和同步,传参为 Map < TopicPartition, OffsetAndMetadata >,其中 key 为消息分区,value 为位移对象。

3. 分组协调者

分组协调者(Group Coordinator)是一个服务,Kafka 集群中的每个节点在启动时都会启动这样一个服务,该服务主要用来存储消费分组相关的元数据信息,每个消费组均会选择一个协调者来负责组内各个分区的消费位移信息存储,选择的主要步骤如下。

① 首先确定消费组的位移信息存入哪个分区。前面提到默认的 __consumer_offsets 主题分区数为 50,通过以下算法可以计算出对应消费组的位移信息应该存入哪个分区:partition = Math.abs(groupId.hashCode() % groupMetadataTopicPartitionCount)。其中 groupId 为消费组的 ID,由消费端指定,groupMetadataTopicPartitionCount 为主题分区数。

② 根据 partition 寻找该分区的 Leader 所对应的节点 Broker,该 Broker 的 Coordinator 即为该消费组的 Coordinator。

4. 重平衡机制

(1) 重平衡发生场景

以下几种场景均会触发重平衡操作。

① 新的消费者加入消费组中。

② 消费者被动下线。例如,消费者长时间的 GC、网络延迟导致消费者长时间未向 Group Coordinator 发送心跳请求,会认为该消费者已经下线并被踢出。

③ 消费者主动退出消费组。

④ 消费组订阅的任意一个主题分区数出现变化。

⑤ 消费组取消某个主题的订阅。

(2) 重平衡操作流程

重平衡的实现可以分为以下几个阶段。

① 查找 Group Coordinator:消费者会从 Kafka 集群中选择一个负载最小的节点发送 GroupCoorinatorRequest 请求,并处理返回响应 GroupCoordinatorResponse。请求参数中包含消费组的 ID,响应中包含 Coordinator 所在节点的 ID、Host 以及端口号信息。

② Join Group:当消费者拿到协调者的信息之后会向协调者发送加入消费组的请求 JoinGroupRequest,当所有的消费者都发送该请求之后,协调者会从中选择一个消费者作为 Leader 角色,然后将组内成员信息、订阅等信息发给消费者,Leader 负责消费方案的分配。

5. 分区分配策略

Kafka 提供了 3 个分区分配策略:RangeAssignor、RoundRobinAssignor 以及 StickyAssignor,下面简单介绍各个算法的实现。

(1) RangeAssignor

Kafka 默认采用此策略进行分区分配,主要流程如下。

假设一个消费组中存在 2 个消费者{C0,C1},该消费组订阅了 3 个主题{T1,T2,T3},每个主题分别存在 3 个分区,一共就有 9 个分区{TP1,TP2,…,TP9}。通过以上算法我们可以得到 D=4,R=1,那么消费者 C0 将消费分区{TP1,TP2,TP3,TP4,TP5},C1 将消费分区{TP6,TP7,TP8,TP9}。这里存在一个问题,如果不能均分,那么前面的消费者将会多消费一个分区。

① 对所有订阅主题下的分区进行排序,得到集合 TP={TP0,TP1,…,TPN+1}。

② 对消费组中的所有消费者根据名字进行字典排序,得到集合 CG={C0,C1,…,CM+1}。

③ 计算 D=N/M,R=N%M。

④ 消费者 C_i 获取的消费分区起始位置=$D*i+\min(i,R)$,C_i 获取的分区总数=D+(if (i+1>R)0 else 1)。

(2) RoundRobinAssignor

使用该策略需要满足两个条件:第一个是消费组中的所有消费者订阅主题应该相同;第二个是同一个消费组中的所有消费者在实例化时给每个主题指定相同的流数。

① 对所有主题的所有分区根据主题+分区得到的哈希值进行排序。

② 对所有消费者按字典排序。

③ 通过轮询的方式将分区分配给消费者。

(3) StickyAssignor

该分配方式在 0.11 版本中开始引入,主要是保证以下特性。

① 尽可能地保证分配均衡。

② 当重新分配时,保留尽可能多的现有分配。

其中第一条的优先级要大于第二条的优先级。

本 章 小 结

① Kafka 是一个高吞吐量、分布式的发布-订阅消息系统。据 Kafka 官网介绍,当前的 Kafka 已经定位为一个分布式流式处理平台(a distributed streaming platform),其以可水平扩展和具有高吞吐量等特性而著称。越来越多的开源分布式处理系统(Flume、Apache Storm、Spark、Flink 等)支持与 Kafka 集成。Kafka 是一款开源的、轻量级的、分布式的、可分区和具有复制备份的、基于 ZooKeeper 协调管理的分布式流平台的、功能强大的消息系统。与传统的消息系统相比,Kafka 能够很好地处理活跃的流数据,使得数据在各个子系统中高性能、低延迟地不停流转。

② 围绕 Kafka 的特性,本章详细介绍了其原理实现,通过主题与日志的深入剖析,说明了 Kafka 内部消息的存放、检索以及删除机制。副本系统中 ISR 概念的引入解决了同步副本与异步副本两种方案各自的缺陷,leader epoch 机制解决了数据丢失以及数据不一致问题。生产端的分区选择算法实现了数据均衡,幂等特性的支持则解决了之前存在的重复消息问题。

③ 本章最后介绍了消费端的相关原理,消费组机制实现了消费端的消息隔离,既有广播也有独占的场景支持,而重平衡机制则保证了消费端的健壮性与扩展性。

第 7 章

Flume 日志处理系统

实际业务场景中,Hadoop 项目的整体开发流程一般如图 7-1 所示。

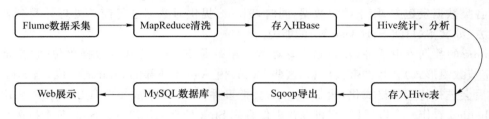

图 7-1　Hadoop 项目的整体开发流程

从 Hadoop 项目的开发流程图中可以看出,在大数据的业务处理过程中,对于数据的采集是十分重要的一步,也是不可避免的一步。

许多公司的平台每天会产生大量的日志(一般为流式数据,如搜索引擎的页面浏览量、查询等),处理这些日志需要特定的日志系统,开源的日志系统包括 Facebook 的 Scribe、Apache 的 Chukwa、LinkedIn 的 Kafka 和 Cloudera 的 Flume 等。一般而言,这些系统需要具有以下特征。

① 构建应用系统和分析系统的桥梁,并将它们之间的关联解耦。
② 支持近实时的在线分析系统和类似于 Hadoop 的离线分析系统。
③ 具有高可扩展性,即当数据量增加时,可以通过增加节点进行水平扩展。

7.1　Flume 的简介

7.1.1　Flume 概述

Flume 作为 Cloudera 开发的实时日志收集系统,受到了业界的认可与广泛应用。Flume 初始的发行版本目前被统称为 Flume OG(Original Generation),属于 Cloudera 公司。

但随着 Flume 功能的扩展,Flume OG 代码工程臃肿、核心组件设计不合理、核心配置不标准等缺点暴露出来,尤其是在 Flume OG 的最后一个发行版本 0.9.4 中,日志传输不稳定的现象尤为严重,为了解决这些问题,2011 年 10 月 22 日,Cloudera 完成了 Flume-728,对 Flume 进行了里程碑式的改动:重构核心组件、核心配置以及代码架构。重构后的版本统称为

Flume NG(Next Generation)。改动的另一个原因是 Flume 被纳入 Apache 旗下,Cloudera Flume 改名为 Apache Flume。

Flume 是 Apache 的顶级项目,官方网站为 http://flume.apache.org/。

Flume 在 0.9.x 和 1.x 之间有较大的架构调整,1.x 版本之后的改称 Flume NG,0.9.x 版本称为 Flume OG。Flume 目前只有 Linux 系统的启动脚本,没有 Windows 系统的启动脚本。

7.1.2 Flume NG 的介绍

1. Flume 特点

Flume 是一个分布式、可靠和高可用的海量日志采集、聚合和传输系统,支持在日志系统中定制各类数据发送方,用于收集数据,同时,Flume 提供对数据进行简单处理并写到各种数据接收方(如文本、HDFS、HBase 等)的能力。

Flume 的数据流由事件(Event)贯穿始终。事件是 Flume 的基本数据单位,它携带日志数据(字节数组形式)并且携带头信息,这些 Event 由 Agent 外部的 Source 生成,当 Source 捕获事件后会进行特定的格式化,然后 Source 会把事件推入(单个或多个)Channel 中。用户可以把 Channel 看作一个缓冲区,它将保存事件直到 Sink 处理完该事件。Sink 负责持久化日志或者把事件推向另一个 Source。

(1) Flume 的可靠性

当节点出现故障时,日志能够被传送到其他节点上而不会丢失。Flume 提供了 3 种级别的可靠性保障,从强到弱依次为:end-to-end(收到数据后 Agent 首先将 Event 写到磁盘上,数据传送成功后再删除;如果数据发送失败,可以重新发送);store on failure(这也是 Scribe 采用的策略,当数据接收方 crash 时,将数据写到本地,待恢复后继续发送);best effort(数据发送到接收方后,不会进行确认)。

(2) Flume 的可恢复性

Flume 的可恢复性是由 Channel 来实现的,Channel 主要包括 File Channel 和 Memory Channel 两种类型,推荐使用 File Channel 将事件持久化到本地文件系统,但是性能上稍差一些。

2. Flume 的一些核心概念

- Client:生产数据,运行在一个独立的线程上。
- Event:一个数据单元,由消息头和消息体组成。Event 可以是日志记录、Avro 对象等。
- Flow:Event 从源点到达目的点的迁移的抽象。
- Agent:一个独立的 Flume 进程,包含组件 Source、Channel、Sink。Agent 使用 JVM 运行 Flume。每台机器运行一个 Agent,但是可以在一个 Agent 中包含多个 Sources 和 Sinks。
- Source:数据收集组件。Source 从 Client 中收集数据,传递给 Channel。
- Channel:中转 Event 的一个临时存储,保存由 Source 组件传递过来的 Event。Channel 连接 Source 和 Sink,有点像一个队列。
- Sink:从 Channel 中读取并移除 Event,将 Event 传递到 FlowPipeline 中的下一个 Agent(如果有的话)。Sink 从 Channel 中收集数据,运行在一个独立的线程上。

3. Flume NG 的体系结构

Flume 运行的核心是 Agent。Flume 以 Agent 为最小的独立运行单位。一个 Agent 就是一个 JVM。Agent 是一个完整的数据收集工具,含有 3 个核心组件,分别是 Source、Channel、Sink。通过这些组件,Event 可以从一个地方流向另一个地方,如图 7-2 所示。

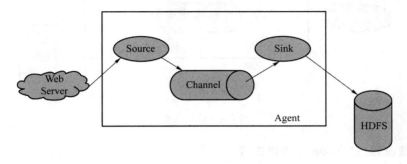

图 7-2 Flume NG 的体系结构

4. Source

Source 是数据的收集端,负责将数据捕获后进行特殊的格式化,将数据封装到事件里,然后将事件推入 Channel 中。

Flume 提供了各种 Source 的实现,包括 Avro Source、Exce Source、Spooling Directory Source、NetCat Source、Syslog Source、Syslog TCP Source、Syslog UDP Source、HTTP Source、HDFS Source 等。如果内置的 Source 无法满足需要,Flume 还支持自定义 Source。

5. Channel

Channel 是连接 Source 和 Sink 的组件,可以将它看作一个数据的缓冲区(数据队列),它可以将事件暂存到内存中,也可以持久化到本地磁盘上,直到 Sink 处理完该事件。

Flume 对于 Channel 则提供了 Memory Channel、JDBC Channel、File Channel 等。

- Memory Channel:可以实现高速的吞吐,但是无法保证数据的完整性。
- JDBC Channel:一般不推荐使用,Event 保存在关系数据库中。
- File Channel:保证数据的完整性与一致性。在具体配置 File Channel 时,建议 File Channel 设置的目录和程序日志文件保存的目录设成不同的磁盘,以便提高效率。

6. Sink

Sink 取出 Channel 中的数据,存入相应的文件系统、数据库,或者提交到远程服务器。

Flume 也提供了各种 Sink 的实现,包括 HDFS Sink、Logger Sink、Avro Sink、File Roll Sink、Null Sink、HBase Sink 等。

Sink 在设置存储数据时,可以向文件系统、数据库、Hadoop 中存储数据。在日志数据较少时,可以将数据存储在文件系统中,并且设定一定的时间间隔来保存数据。在日志数据较多时,可以将相应的日志数据存储到 Hadoop 中,便于日后进行相应的数据分析。

7.1.3 Flume 的部署类型

1. 单一流程

单一流程如图 7-3 所示。

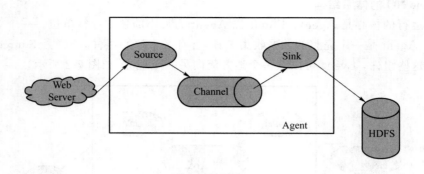

图 7-3　单一流程

2．多代理流程（多个 Agent 顺序连接）

如图 7-4 所示，可以将多个 Agent 顺序连接起来，将最初的数据源经过收集，存储到最终的存储系统中。这是最简单的情况，一般情况下，应该控制这种顺序连接的 Agent 的数量，因为数据流经的路径变长了，如果不考虑 failover（故障转移）的话，出现故障将影响整个 Flow 上的 Agent 收集服务。

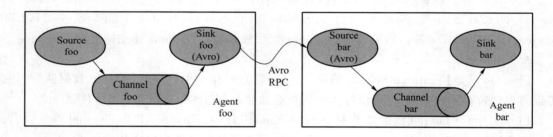

图 7-4　多代理流程

3．流的合并（多个 Agent 的数据汇聚到同一个 Agent 上）

这种情况应用的场景比较多，如要收集 Web 网站的用户行为日志，Web 网站为了提高可用性使用的是负载集群模式，每个节点都会产生用户行为日志，可以为每个节点都配置一个 Agent 来单独收集日志数据，然后多个 Agent 将数据最终汇聚到一个用来存储数据的存储系统（如 HDFS）上，如图 7-5 所示。

4．多路复用流（多级流）

Flume 还支持多级流，什么是多级流？举个例子，当 Syslog、Java、Nginx、Tomcat 等混合在一起的日志流流入一个 Agent 后，可以将 Agent 中混杂的日志流分开，然后给每种日志建立一个自己的传输通道，如图 7-6 所示。

5．负载均衡功能

图 7-7 中，Agent1 是一个路由节点，负责将 Channel 暂存的 Event 均衡到对应的多个 Sink 组件上，而每个 Sink 组件分别连接到一个独立的 Agent 上。

第 7 章 Flume 日志处理系统

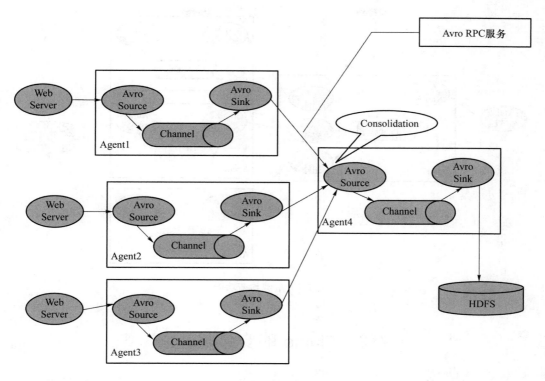

图 7-5 流的合并

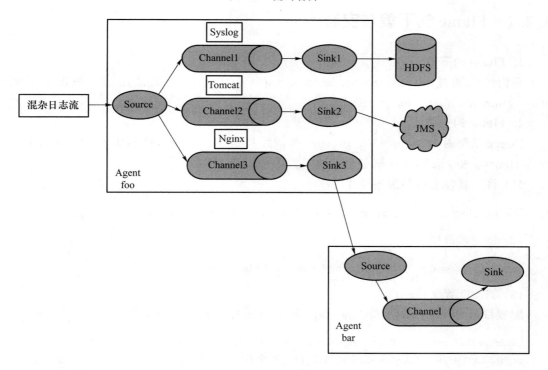

图 7-6 多路复用流

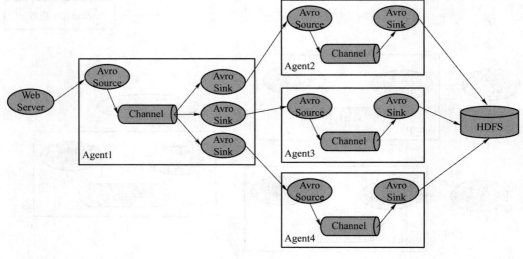

图 7-7 负载均衡功能

7.2 Flume 的安装与配置

7.2.1 Flume 的下载与安装

1. Flume 的下载

我们可以通过 http://apache.hadoop.org 下载 Flume,建议使用国内镜像下载地址 http://mirrors.hust.edu.cn/apache/。

2. Flume 的安装

Flume 框架对 Hadoop 和 ZooKeeper 的依赖只是在 jar 包上,并不要求 Flume 启动时必须将 Hadoop 和 ZooKeeper 服务也启动。

(1) 将安装包上传到服务器并解压

[hadoop@hadoop1 ~]$ tar -zxvf apache-flume-1.8.0-bin.tar.gz -C apps/

(2) 创建软链接

[hadoop@hadoop1 ~]$ ln -s apache-flume-1.8.0-bin/ flume

(3) 修改配置文件

配置目录为/home/hadoop/apps/apache-flume-1.8.0-bin/conf。

[hadoop@hadoop1 conf]$ cp flume-env.sh.template flume-env.sh

export JAVA_HOME=/usr/local/jdk1.8.0_73

(4) 配置环境变量

[hadoop@hadoop1 conf]$ vi ~/.bashrc

```
export FLUME_HOME=/home/hadoop/apps/flume
export PATH=$PATH:$FLUME_HOME/bin
```

保存使其立即生效：

```
[hadoop@hadoop1 conf]$ source ~/.bashrc
```

(5) 查看版本

```
[hadoop@hadoop1 ~]$ flume-ng version
Flume 1.8.0
Source code repository: https://git-wip-us.apache.org/repos/asf/flume.git
Revision: 2561a23240a71ba20bf288c7c2cda88f443c2080
Compiled by denes on Mon May 4 11:15:44 PDT 2020
From source with checksum b29e416802ce9ece3269d34233baf43f
```

7.2.2 Flume Sources 描述

1. Avro Source

监听 Avro 端口，从 Avro Client Streams 接收 Events。当与另一个（前一跳）Flume Agent 内置的 Avro Sink 配对时，它可以创建分层收集拓扑。

参考配置示例，实现 Avro Source：

```
#配置一个 Agent, Agent 的名称可以自定义（如 a1）
#分别指定 Agent 的 Sources、Sinks、Channels 的名称, 名称可以自定义
a1.sources = s1
a1.sinks = k1
a1.channels = c1
#配置 Sources
a1.sources.s1.channels = c1
a1.sources.s1.type = avro
a1.sources.s1.bind = 192.168.123.102
a1.sources.s1.port = 6666
#配置 Channels
a1.channels.c1.type = memory
#配置 Sinks
a1.sinks.k1.channel = c1
a1.sinks.k1.type = logger
#为 Sources 和 Sinks 绑定 Channels
a1.sources.s1.channels = c1
a1.sinks.k1.channel = c1
#启动 Flume
[hadoop@hadoop1 ~]$ flume-ng agent --conf conf --conf-file ~/apps/flume/examples/single_avro.properties --name a1 -Dflume.root.logger=DEBUG,console -Dorg.apache.flume.log.printconfig=true -Dorg.apache.flume.log.rawdata=true
```

通过 Flume 提供的 Avro 客户端向指定机器的指定端口发送日志信息：

```
[hadoop@hadoop1 ~]$ flume-ng avro-client -c ~/apps/flume/conf -H 192.168.123.102 -p 8888 -F file1.txt
```

2. Thrift Source

Thrift Source 与 Avro Source 基本一致。只要把 Source 的类型改成 Thrift 即可，如 a1.sources.r1.type = thrift，比较简单，不做赘述。

3. Exec Source

Exec Source 的配置就是设定一个 UNIX(Linux)命令，然后通过这个命令不断输出数据。如果进程退出，Exec Source 也一起退出，不会产生进一步的数据。

参考配置示例，实现 Exec Source：

```
# 配置一个 Agent，命名为 a1
a1.sources = s1
a1.sinks = k1
a1.channels = c1
# 配置 Sources
a1.sources.s1.type = exec
a1.sources.s1.command = tail -F /home/hadoop/logs/test.log
a1.sources.s1.channels = c1
# 配置 Sinks
a1.sinks.k1.type = logger
a1.sinks.k1.channel = c1
# 配置 Channels
a1.channels.c1.type = memory
```

使用配置文件启动 Flume 日志服务：

```
[hadoop@hadoop1 ~]$ flume-ng agent --conf conf --conf-file ~/apps/flume/examples/case_exec.properties --name a1 -Dflume.root.logger=DEBUG,console -Dorg.apache.flume.log.printconfig=true -Dorg.apache.flume.log.rawdata=true
```

4. JMS Source

从 JMS(Java 消息服务)系统(消息、主题)中读取数据。

5. Spooling Directory Source

Spooling Directory Source 监测配置的目录下新增的文件，并将文件中的数据读取出来。需要注意的是：①复制到 spool 目录下的文件不可以再打开编辑；②spool 目录下不可包含相应的子目录。Spooling Directory Source 主要用于对日志的准实时监控。

参考配置示例，实现 Spooling Directory Source：

```
a1.sources = s1
a1.sinks = k1
a1.channels = c1
# 配置 Sources
a1.sources.s1.type = spooldir
a1.sources.s1.spoolDir = /home/hadoop/logs
a1.sources.s1.fileHeader = true
```

```
a1.sources.s1.channels = c1
# 配置 Sinks
a1.sinks.k1.type = logger
a1.sinks.k1.channel = c1
# 配置 Channels
a1.channels.c1.type = memory
```

启动命令：

[hadoop@hadoop1 ~]$ flume-ng agent --conf conf --conf-file /home/hadoop/apps/flume/examples/case_spool.properties --name a1 -Dflume.root.logger = INFO,console

7.3 Flume 代理流配置

7.3.1 单一代理流配置

1. 官网介绍

参考 http://flume.apache.org/FlumeUserGuide.html#avro-source 网址，通过一个通道连接源和接收器。需要为指定的代理配置源、接收器和通道。一个源的实例可以指定多个通道，但只能指定一个接收器实例。

2. 配置单个组件

定义流之后，需要设置每个源、接收器和通道的属性。可以分别设定组件的属性值。

7.3.2 单代理多流配置

单个 Flume 代理可以包含几个独立的流。用户可以在一个配置文件中列出多个源、接收器和通道，这些组件可以连接形成多个流。

可以连接源和接收器到其相应的通道，设置两个不同的流。例如，如果需要设置一个 agent_foo 代理两个流，一个是从外部 Avro 客户端到 HDFS，另一个是 tail 的输出到 Avro 接收器，可以在这里做配置。

7.3.3 配置多代理流程

设置一个多层的流，需要有一个指向下一跳 Avro 源的第一跳的 Avro 接收器。这将导致第一个 Flume 代理转发事件到下一个 Flume 代理。例如，如果定期发送文件，每个事件（一个文件）Avro 客户端使用本地 Flume 代理，那么这个本地的代理可以转发到另一个有存储的代理。

7.3.4 多路复用流

Flume 支持扇出流（从一个源到多个通道）。有两种模式的扇出：复制和复用。在复制的情况下,事件被发送到所有的配置通道。在复用的情况下,事件被发送到合格的渠道,只有一个子集。扇出流需要指定源和扇出通道的规则,也就是添加一个通道"选择",可以是复制或复用,再进一步指定选择的规则(如果它是一个多路),默认情况下为复制。

本章小结

① Flume 是 Cloudera 提供的一个分布式、可靠、可用的系统,它能够对不同数据源的海量日志数据进行高效收集、聚合、移动,最后存储到一个中心化数据存储系统中。由原来的 Flume OG 到现在的 Flume NG,Flume 进行了架构重构,并且现在的 NG 版本完全不兼容原来的 OG 版本。经过架构重构后,Flume NG 更像是一个轻量级的小工具,非常简单,容易适应各种方式的日志收集,并支持故障转移和负载均衡。

② Flume 由 Source、Channel、Sink 组成。

第 8 章 ZooKeeper 分布式协调系统

Hadoop 集群中大量节点的配置信息如何做到全局一致并且使单点修改迅速响应到整个集群？如何更有效地实现 Hadoop 集群配置管理？

Hadoop 集群中的 NameNode 和 ResourceManager 的单点故障怎么解决？重点是集群的主节点的单点故障。

8.1 分布式协调技术概述

1. 分布式协调技术

在介绍 ZooKeeper 组件之前先给大家介绍一种技术——分布式协调技术。什么是分布式协调技术？其实分布式协调技术主要用来解决分布式环境中多个进程之间的同步控制，让它们有序地去访问某种临界资源，防止造成"脏数据"的后果。这时有人可能会说，这个简单，写一个调度算法就轻松解决了。说这句话的人可能对分布式系统不是很了解，所以才会出现这种误解。如果这些进程全部跑在一台机器上，相对来说确实就好办了，问题就在于是在一个分布式的环境下，这时候问题又来了，什么是分布式呢？下面通过一张图来帮助大家理解这方面的内容，如图 8-1 所示。

给大家分析一下这张图。在图 8-1 中有 3 台服务器，每台服务器各自运行一个应用程序。然后我们将这 3 台机器通过网络连接起来，构成一个系统来为用户提供服务，对用户来说，这个系统的架构是透明的，用户感觉不到这个应用系统是一个什么样的架构，那么我们就可以把这种系统称作分布式系统。

接下来分析在这个分布式系统中如何对服务进程进行调度，假设在第一台机器上挂载了一个资源，然后这 3 个物理分布的进程都要竞争这个资源，但我们又不希望它们同时进行访问，这时候就需要一个协调器来让它们有序地访问这个资源。这个协调器就是我们经常提到的"锁"。例如，进程 1 在使用该资源的时候，会先去获得锁，进程 1 获得锁以后会对该资源保持独占，这样其他进程就无法访问该资源，进程 1 用完该资源以后就将锁释放掉，让其他进程来获得锁。那么通过这个锁机制，我们就能保证分布式系统中多个进程能够有序地访问该临界资源。我们把分布式环境下的这个锁称为分布式锁。分布式锁也就是分布式协调技术实现的核心内容。

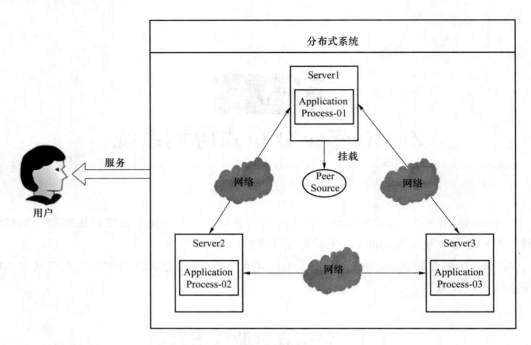

图 8-1 分布式协调机制

2. 分布式锁的实现

在了解了分布式环境之后,有人可能会感觉这不是很难。无非是将原来在同一台机器上对进程进行调度的原语通过网络实现在分布式环境中。表面上的确可以这么说。但是问题就在网络这里,在分布式系统中,所有在同一台机器上的假设都不存在,因为网络是不可靠的。

在同一台机器上,对一个服务的调用如果成功,那就是调用成功,如果失败,如抛出异常,那就是调用失败。但是在分布式环境中,由于网络的不可靠,对一个服务的调用失败了并不表示一定是失败的,可能是执行成功了,但是响应返回的时候失败了。还有一些情况,如 A 和 B 都去调用 C 服务,在时间上 A 先调用,然后 B 再调用,那么最后的结果一定是 A 的请求先于 B 到达吗?这些在同一台机器上的种种假设,我们都要重新思考,我们还要思考这些问题给我们的设计和编码带来了哪些影响。在分布式环境中,为了提升可靠性,我们往往会部署多套服务,但是如何在多套服务中达到一致性?在同一台机器上多个进程之间的同步相对来说比较容易办到,但在分布式环境中却是一个大难题。

所以分布式协调比在同一台机器上对多个进程的调度要难得多,而且如果为每一个分布式应用都开发一个独立的协调程序,一方面,协调程序的反复编写浪费时间和资源,且难以形成通用、伸缩性好的协调器,另一方面,协调程序开销比较大,会影响系统原有的性能。所以急需一种高可靠、高可用的通用协调机制,用以协调分布式应用。

3. 分布式锁的实现者

目前在分布式协调技术方面做得比较好的就是 Google 的 Chubby 和 Apache 的 ZooKeeper,它们都是分布式锁的实现者。有人会问,既然有了 Chubby,为什么还要开发 ZooKeeper,难道 Chubby 做得不够好吗?不是这样的,主要是因为 Chubby 是非开源的。后来雅虎模仿 Chubby 开发出了 ZooKeeper,实现了类似的分布式锁功能,并且将 ZooKeeper 作为一种开源的程序捐献给了 Apache,这样用户就可以使用 ZooKeeper 所提供的锁服务。

ZooKeeper 在分布式领域久经考验,它的可靠性、可用性都是经过理论和实践的验证的。所以我们在构建一些分布式系统的时候,就可以以这类系统为起点来构建我们的系统,这将节省不少成本,而且 bug 也将更少。

4. CAP 理论

CAP 理论指出,对于一个分布式计算系统来说,不可能同时满足以下 3 点。

① 一致性:在分布式环境中,一致性是指数据在多个副本之间能够保持一致的特性,等同于所有节点访问同一份最新的数据副本。在一致性的需求下,当一个系统在数据一致的状态下执行更新操作后,应该保证系统的数据仍然处于一致的状态。

② 可用性:每次请求都能获取到正确的响应,但是不保证获取的数据为最新数据。

③ 分区容错性:分布式系统在遇到任何网络分区故障的时候,仍然需要能够保证对外提供满足一致性和可用性的服务,除非整个网络环境都发生了故障。

一个分布式系统最多只能同时满足一致性(Consistency)、可用性(Availability)和分区容错性(Partition Tolerance)这 3 项中的 2 项,如图 8-2 所示。在这 3 个基本需求中,P 是必需的,因此只能在 CP 和 AP 中选择,ZooKeeper 保证的是 CP,Spring Cloud 系统中的注册中心 Eruka 实现的是 AP。

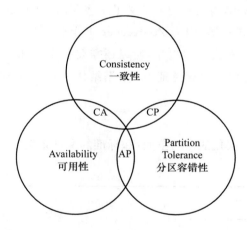

图 8-2 CAP 理论

5. BASE 理论

BASE 是 Basically Available(基本可用)、Soft-state(软状态)和 Eventually Consistent(最终一致性)3 个短语的缩写。

- 基本可用:在分布式系统出现故障时,允许损失部分可用性(服务降级、页面降级)。
- 软状态:允许分布式系统出现中间状态。而且中间状态不影响系统的可用性。这里的中间状态是指不同的 Data Replication(数据备份节点)之间的数据更新可以出现延时的最终一致性。
- 最终一致性:Data Replications 经过一段时间达到一致性。

BASE 理论是对 CAP 理论中的一致性和可用性进行权衡的结果,BASE 理论的核心思想就是:我们无法做到强一致,但每个应用都可以根据自身的业务特点,采用适当的方式来使系统达到最终一致性。

8.2 ZooKeeper 概述

ZooKeeper 是一种开放源码的分布式应用程序协调服务，是 Google 的 Chubby 组件的一个开源实现。它提供了简单原始的功能，分布式应用可以基于它实现更高级的服务，如分布式同步、配置管理、集群管理、命名管理、队列管理。它被设计为易于编程，使用文件系统目录树作为数据模型。服务端运行在 Java 环境中，提供 Java 和 C 的客户端 API。众所周知，协调服务非常容易出错，但是却很难恢复正常，例如，协调服务很容易处于竞争状态以至于出现死锁。设计 ZooKeeper 的目的是减轻分布式应用程序所承担的协调任务，ZooKeeper 是集群的管理者，监视着集群中各节点的状态，根据节点提交的反馈进行下一步合理的操作，最终将简单易用的接口和功能稳定、性能高效的系统提供给用户。

前面提到了那么多的服务，如分布式锁、CAP 理论、BASE 理论等，它们是如何实现的？相信这才是大家关心的问题。ZooKeeper 在实现这些服务时，首先设计了一种新的数据结构——Znode，然后在该数据结构的基础上定义了一些原语，也就是关于该数据结构的一些操作。有了数据结构和原语还不够，因为 ZooKeeper 工作在一个分布式的环境下，服务是通过消息以网络的形式发送给分布式应用程序，所以还需要一个通知机制——Watch 机制。总结一下，ZooKeeper 所提供的服务主要是通过"数据结构＋原语＋Watch 机制"3 个部分来实现的。

1. Znode

ZooKeeper 拥有一个层次化的命名空间，和标准的文件系统非常相似，如图 8-3 所示。

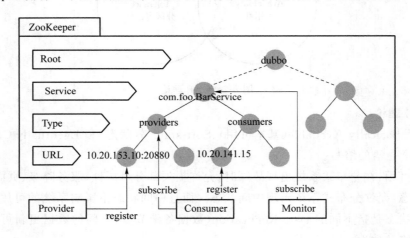

图 8-3 ZooKeeper 的数据模型

从图 8-3 中我们可以看出，ZooKeeper 的数据模型在结构上和标准文件系统的非常相似，都是采用树形层次结构，ZooKeeper 树中的每个节点被称为 Znode。和文件系统的目录树一样，ZooKeeper 树中的每个节点可以拥有子节点。但也有不同之处，主要是以下几点。

(1) 引用方式

Znode 通过路径引用,如同 UNIX 中的文件路径。路径必须是绝对的,因此它们必须由斜杠字符来开头。除此以外,路径必须是唯一的,也就是说,每个路径只有一种表示,因此这些路径不能改变。在 ZooKeeper 中,路径由 Unicode 字符串组成,并且有一些限制。字符串"/zookeeper"用以保存管理信息,如关键配额信息。

(2) Znode 结构

ZooKeeper 命名空间中的 Znode 兼具文件和目录两种特点,既像文件一样维护着数据、元信息、访问控制列表(ACL)、时间戳等数据结构,又像目录一样可以作为路径标识的一部分。每个 Znode 由以下三部分组成。

- stat:此为状态信息,描述该 Znode 的版本、权限等信息。
- data:与该 Znode 关联的数据。
- children:该 Znode 下的子节点。

ZooKeeper 虽然可以关联一些数据,但并没有被设计为常规的数据库或者大数据存储,它用来管理调度数据,如分布式应用中的配置文件信息、状态信息、汇集位置等。这些数据的共同特性就是它们都是很小的数据,通常以千字节为大小单位。ZooKeeper 的服务器和客户端都被设计为严格检查并限制每个 Znode 的数据大小至多为 1 MB,但常规使用中应该远小于此值。

(3) 数据访问

ZooKeeper 中每个节点存储的数据要被原子性地操作。也就是说,读操作将获取与节点相关的所有数据,写操作将替换节点的所有数据。另外,每一个节点都拥有自己的 ACL,这个列表规定了用户的权限,即限定了特定用户对目标节点可以执行的操作。

(4) 节点类型

ZooKeeper 中有两种节点,分别为临时节点和永久节点。节点的类型在创建时即被确定,并且不能改变。

- 临时节点:该节点的生命周期依赖于创建它们的会话(Session)。一旦会话结束,临时节点将被自动删除,当然也可以手动删除。虽然每个临时的 Znode 都会绑定到一个客户端会话,但它们对所有的客户端还是可见的。另外,ZooKeeper 的临时节点不允许拥有子节点。
- 永久节点:该节点的生命周期不依赖于会话,并且只有在客户端显示执行删除操作的时候,它们才能被删除。

(5) 顺序节点

当创建 Znode 的时候,用户可以请求在 ZooKeeper 的路径结尾添加一个递增的计数。这个计数对于此节点的父节点来说是唯一的,它的格式为"％10d"(10 位数字,没有数值的数位用 0 补齐,如"0000000001")。当计数值大于 $2^{32}-1$ 时,计数器将溢出。

(6) 观察

客户端可以在节点上设置 Watch,我们称之为监视器。当节点状态发生改变时,Znode 的增、删、改等操作将会触发 Watch 所对应的操作。当 Watch 被触发时,ZooKeeper 将会向客户端发送且仅发送一条通知,因为 Watch 只能被触发一次,这样可以减少网络流量。

2. ZooKeeper 中的时间

(1) 时间戳

致使 ZooKeeper 节点状态改变的每一个操作都将使节点接收到一个 Zxid 格式的时间戳,

并且这个时间戳全局有序。也就是说,每个对节点的改变都将产生一个唯一的 Zxid。如果 Zxid1 的值小于 Zxid2 的值,那么 Zxid1 所对应的事件发生在 Zxid2 所对应的事件之前。实际上,ZooKeeper 的每个节点维护着 3 个 Zxid 值,如下。
- cZxid:节点的创建时间所对应的 Zxid 格式时间戳。
- mZxid:节点的修改时间所对应的 Zxid 格式时间戳。
- pZxid:与该节点的子节点(或该节点)的最近一次创建/删除的时间戳对应。

实现中 Zxid 是一个 64 位的数字,其高 32 位是 epoch,用来标识 Leader 关系是否改变,每当一个 Leader 被选出来,它都会有一个新的 epoch,低 32 位是个递增计数。

(2) 版本号

对节点的每一个操作都将使这个节点的版本号增加。每个节点维护着 3 个版本号,如下。
- dataVersion:节点数据版本号。
- cVersion:子节点版本号。
- aclVersion:节点所拥有的 ACL 版本号。

3. ZooKeeper 节点属性

一个节点自身拥有表示其状态的许多重要属性,如表 8-1 所示。

表 8-1 ZooKeeper 节点属性

状态属性	说明
cZxid	数据节点创建时的事务 ID
ctime	数据节点创建的时间
mZxid	数据节点最后一次更新时的事务 ID
mtime	数据节点最后一次更新的时间
pZxid	数据节点的子节点列表最后一次被修改(是子节点列表变更,而不是子节点内容变更)时的事务 ID
cVersion	子节点的版本号,当添加/删除子节点的时候发生改变
dataVersion	数据节点的版本号,仅当 data 属性发生改变的时候才改变
aclVersion	数据节点的 ACL 版本号
ephemeralOwner	如果节点是临时节点,则表示创建该节点的会话的 SessionID;如果节点是永久节点,则该属性值为 0
dataLength	数据内容的长度
numChildren	数据节点当前的子节点个数

4. ZooKeeper 中的 API 操作

ZooKeeper 中有 9 个基本 API 操作,如表 8-2 所示。

表 8-2 ZooKeeper 中的 API 操作

功能	描述
create	在本地目录树中创建一个节点
delete	删除一个节点
exists	测试本地是否存在目标节点

功能	描述
getData/setData	从目标节点上读取／写数据
getACL/setACL	获取／设置目标节点访问控制列表信息
getChildren	检索一个子节点上的列表
sync	等待要被传送的数据

更新 ZooKeeper 操作是有限制的，delete 或 setData 必须明确要更新的 Znode 的版本号，可以通过调用 exists 找到，如果版本号不匹配，更新将会失败。

更新 ZooKeeper 操作是非阻塞式的。如果另一个进程在同时更新这个 Znode，导致客户端失去了一个更新，它可以在不阻塞其他进程执行的情况下，选择重新尝试或进行其他操作。

尽管 ZooKeeper 可以被看作一个文件系统，但是出于便利，它摒弃了一些文件系统的操作原语。因为文件非常小并且是整体读写的，所以不需要打开、关闭等操作。

8.3 ZooKeeper 监听机制

8.3.1 Watch 触发器

1. Watch 概述

ZooKeeper 可以为所有的读操作设置 Watch，这些读操作包括 exists、getChildren 和 getData。Watch 事件是一次性的触发器，当 Watch 的对象状态发生改变时，将会触发此对象上 Watch 所对应的事件。Watch 事件将被异步地发送给客户端，并且 ZooKeeper 为 Watch 机制提供了有序的一致性保证。理论上，客户端接收 Watch 事件的时间要快于其看到 Watch 对象状态变化的时间。

2. Watch 类型

ZooKeeper 所管理的 Watch 可以分为以下两类。

- 数据 Watch(Data Watch)：getData 和 exists 负责设置数据 Watch。
- 孩子 Watch(Child Watch)：getChildren 负责设置孩子 Watch。

我们可以通过操作返回的数据来设置不同的 Watch，如下。

- getData 和 exists：返回关于节点的数据信息。
- getChildren：返回孩子列表。

因此具有以下几个特征。

- 一个成功的 setData 操作将触发 Znode 的数据 Watch。
- 一个成功的 create 操作将触发 Znode 的数据 Watch 以及孩子 Watch。
- 一个成功的 delete 操作将触发 Znode 的数据 Watch 以及孩子 Watch。

3. Watch 注册与触发

Watch 由客户端所连接的 ZooKeeper 服务器在本地维护，因此 Watch 可以非常容易地设置、管理和分派。当客户端连接到一个新的服务器时，任何的会话事件都有可能触发 Watch。

另外，当从服务器断开连接的时候，Watch 将不会被接收。但是，当一个客户端重新建立连接的时候，任何之前注册过的 Watch 都会被重新注册。

一般情况下，会有如下的场景操作。

- exists 操作上的 Watch 在被监视的 Znode 被创建、被删除或数据更新时被触发。
- getData 操作上的 Watch 在被监视的 Znode 被删除或数据更新时被触发。在被创建时不能被触发，因为只有 Znode 一定存在，getData 操作才会成功。
- getChildren 操作上的 Watch 在被监视的 Znode 的子节点被创建或删除，或是这个 Znode 自身被删除时被触发。可以通过查看 Watch 事件类型来区分是 Znode 被删除还是它的子节点被删除：NodeDelete 表示 Znode 被删除，NodeDeletedChanged 表示子节点被删除。

8.3.2 监听原理

ZooKeeper 的 Watch 机制主要包括客户端线程、客户端 WatchManager、ZooKeeper 服务器三部分。WatchManager 类用于管理 Watch 和相应的触发器。客户端在向 ZooKeeper 服务器注册的同时，会将 Watch 对象存储在客户端的 WatchManager 中。ZooKeeper 服务器触发 Watch 事件后，会向客户端发送通知，客户端线程从 WatchManager 中取出对应的 Watch 对象来执行回调逻辑。ZooKeeper 监听原理可以参考图 8-4。

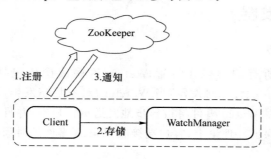

图 8-4　ZooKeeper 监听原理

8.3.3 ZooKeeper 应用举例

1. 分布式锁应用场景

在分布式锁服务中，有一种最典型的应用场景，就是通过对集群进行 Master 选举来解决分布式系统中的单点故障。分布式系统中的单点故障是：通常分布式系统采用主从模式，就是一个主控机连接多个处理节点，主节点负责分发任务，从节点负责处理任务，当主节点发生故障时，整个系统就都瘫痪了，我们把这种故障称为单点故障。如图 8-5 和图 8-6 所示。

2. 传统解决方案

传统解决方案是采用一个备用节点，这个备用节点定期向当前主节点发送 Ping 包，主节点收到 Ping 包后向备用节点发送回复 Ack，当备用节点收到回复的时候就认为当前主节点还活着，让它继续提供服务，如图 8-7 所示。

当主节点"挂了"，这时候备用节点收不到回复了，然后备用节点就认为主节点"挂了"而接

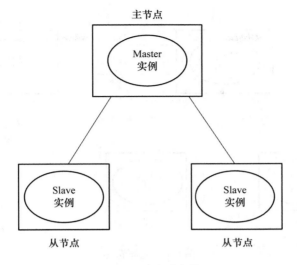

图 8-5　分布式架构

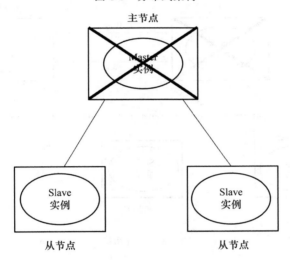

图 8-6　单点故障

替它成为主节点,如图 8-8 所示。

但是这种方式有一个隐患,就是网络故障,网络故障带来的问题如图 8-9 所示。

也就是说,主节点并没有挂掉,只是在回复的时候网络发生故障,这样一来,备用节点收不到回复,就会认为主节点"挂了",然后备用节点将它的 Master 实例启动起来,这样分布式系统中就有了两个主节点,也就是双 Master,出现双 Master 以后,从节点就会将它所做的事一部分汇报给主节点,一部分汇报给备用节点,这样服务就全乱了。为了防止出现这种情况,我们引入了 ZooKeeper 分布式协调机制,它虽然不能避免网络故障,但能够保证每时每刻只有一个 Master。

3. ZooKeeper 解决方案

(1) Master 启动

在引入 ZooKeeper 以后,我们启动了两个主节点——主节点-A 和主节点-B,它们启动以后,都向 ZooKeeper 注册一个节点。我们假设主节点-A 注册的节点是 master-00001,主节点-B 注册的节点是 master-00002,注册完以后进行选举,编号最小的节点将在选举中获胜,获得锁成

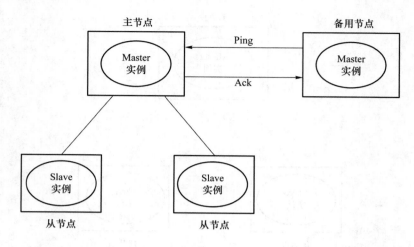

图 8-7 主备节点

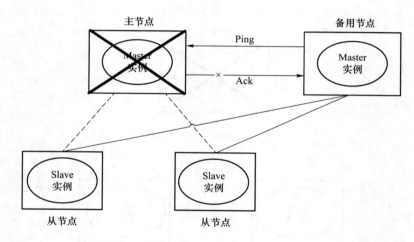

图 8-8 主备切换

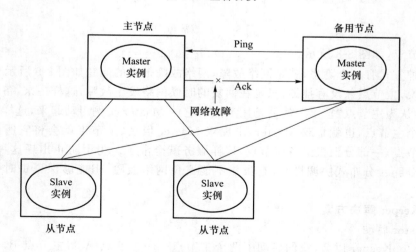

图 8-9 网络故障带来的问题

为主节点,也就是说,主节点-A 将会获得锁成为主节点,然后主节点-B 将被阻塞为一个备用节点,如图 8-10 所示。那么,通过这种方式就完成了对两个 Master 进程的调度。

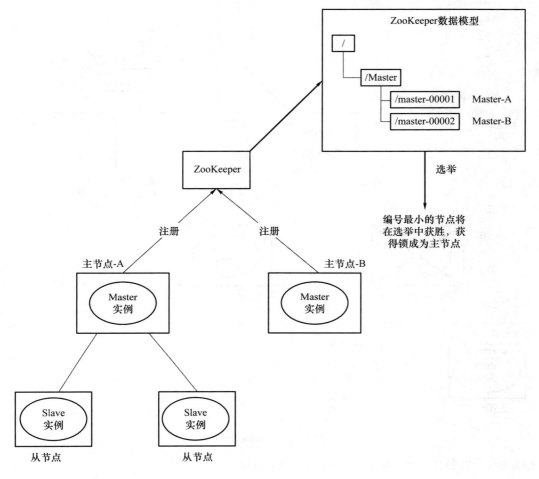

图 8-10　Master 启动

（2）Master 故障

如果主节点-A"挂了",它所注册的节点将被自动删除,ZooKeeper 会自动感知节点的变化,然后再次发起选举,这时候主节点-B 将在选举中获胜,替代主节点-A 成为主节点,如图 8-11 所示。

（3）Master 恢复

如果主节点-A 恢复了,它会再次向 ZooKeeper 注册一个节点,这时候它注册的节点会是 master-00003,ZooKeeper 会自动感知节点的变化,再次发起选举,主节点-B 在选举中会再次获胜,继续担任主节点,主节点-A 会担任备用节点,如图 8-12 所示。

8.4　ZooKeeper 的安装与集群配置

鉴于 ZooKeeper 集群本身的特点,服务器集群的节点数推荐设置为奇数。这里规划为 3

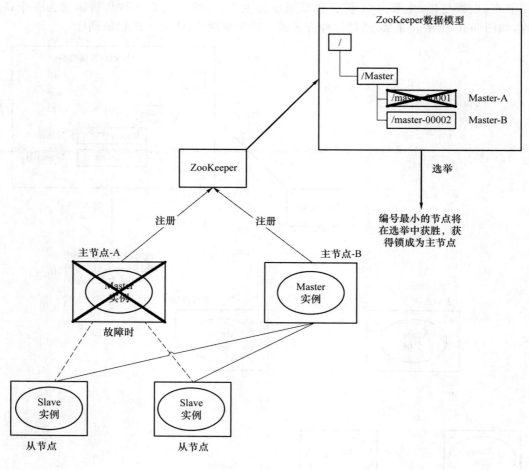

图 8-11 Master 故障

台服务器，主机名分别为 hadoop1、hadoop2、hadoop3。

8.4.1 ZooKeeper 的安装

1. ZooKeeper 的下载

推荐下载地址：http://mirrors.hust.edu.cn/apache/ZooKeeper/。本书使用的是 3.4.10 版本。

解压安装到自己的目录：

```
[hadoop@hadoop1 ~]$ tar -zxvf zookeeper-3.4.10.tar.gz -C apps/
```

2. 修改配置文件

```
[hadoop@hadoop1 zookeeper-3.4.10]$ cd conf/
[hadoop@hadoop1 conf]$ mv zoo_sample.cfg zoo.cfg
[hadoop@hadoop1 conf]$ vim zoo.cfg
```

修改数据目录字段：

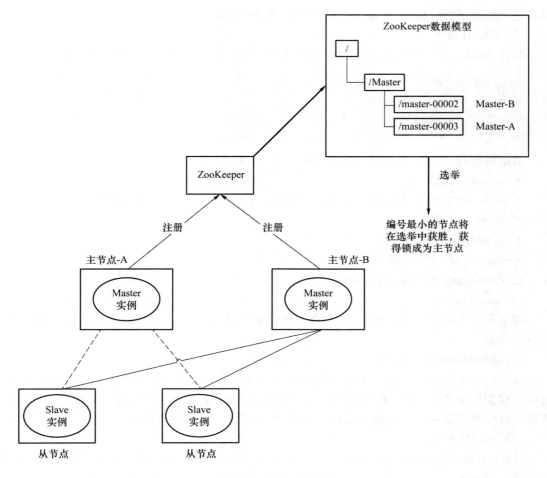

图 8-12　Master 恢复

```
dataDir = /home/hadoop/data/zkdata/
```

3．ZooKeeper 的基本配置参数

（1）tickTime

心跳基本时间，单位为毫秒，ZooKeeper 基本上所有的时间都是这个时间的整数倍。

（2）initLimit

tickTime 的个数，表示在 Leader 选举结束后，Followers 与 Leader 同步需要的时间，如果 Followers 比较多或者 Leader 的数据非常多，同步时间可能会相应增加，那么这个参数的值也需要相应增加。当然，这个参数也是 Follower 和 Observer 在开始同步 Leader 的数据时的最大等待时间（setSoTimeout）。

（3）syncLimit

tickTime 的个数，这个参数容易和 initLimit 混淆，它也表示 Follower 和 Observer 与 Leader 交互时的最大等待时间，只不过是在与 Leader 同步完毕之后，进入正常请求转发或 Ping 等消息交互时的最大等待时间。

（4）dataDir

内存数据库快照存放地址，如果没有指定事务日志存放地址（dataLogDir），其默认也是存放在这个路径下,建议两个地址分开存放到不同的设备上。

(5) clientPort

配置 ZooKeeper 监听客户端连接的端口,在配置文件 zoo.cfg 末尾添加如下字段:

```
dataLogDir = /home/hadoop/log/zklog/
server.1 = hadoop1:2888:3888
server.2 = hadoop2:2888:3888
server.3 = hadoop3:2888:3888
```

语法结构:server.serverid = host:tickpot:electionport。
- server:固定写法。
- serverid:每台服务器的指定 ID(必须处于 1~255 之间,每一台机器不能重复)。
- host:主机名。
- tickport:心跳通信端口。
- electionport:选举端口。

4. ZooKeeper 的高级配置参数

(1) dataLogDir

将事务日志存储在该路径下,这个参数比较重要,日志存储的设备效率会影响 ZooKeeper 的写吞吐量。

(2) globalOutstandingLimit

(Java system property:zookeeper.globalOutstandingLimit)默认值是 1 000,限定了所有已经连接到服务器上,但是还没有返回响应的请求个数(即所有客户端请求的总数,不是连接总数),这个参数是针对单台服务器而言,设定太大可能会导致内存溢出。

(3) preAllocSize

(Java system property:zookeeper.preAllocSize)默认值是 64 MB,以千字节为单位,预先分配额定空间用于后续 transactionlog 写入,每当剩余空间小于 4 KB 时,就会再分配 64 MB,如此循环。如果 SNAP 做得比较频繁(snapCount 比较小的时候),那么适当减少这个值。

(4) snapCount

(Java system property:zookeeper.snapCount)默认值是 100 000,transaction 每达到 snapCount/2 + rand.nextInt(snapCount/2)时,就做一次 Snapshot(快照),默认情况下是 50 000~100 000 条 transactionlog 做一次,采用随机数是为了避免所有服务器在同一时间做 Snapshot。

(5) traceFile

(Java system property:zookeeper.traceFile)ZooKeeper 使用 log4j 来记录系统日志。默认情况下,系统日志文件由 log4j 配置文件来决定配置。

(6) maxClientCnxns

默认值是 10,一个客户端能够连接到同一台服务器上的最大连接数,根据 IP 来区分。如果设置为 0,表示没有任何限制。设置该值一方面是为了防止 DoS 攻击。

(7) clientPortAddress

与 clientPort 匹配,表示某个 IP 地址,如果服务器有多个网络接口(多个 IP 地址),而没有设置这个属性,则 clientPort 会绑定到所有 IP 地址上,否则只绑定到设置的 IP 地址上。

(8) minSessionTimeout

最小的 sessionTimeout 时间,默认值是 2 个 tickTime,客户端设置的 sessionTimeout 如

果小于这个值,则会被强制协调为这个最小值。

(9) maxSessionTimeout

最大的 sessionTimeout 时间,默认值是 20 个 tickTime,客户端设置的 sessionTimeout 如果大于这个值,则会被强制协调为这个最大值。

(10) electionAlg

领导选举算法,默认值是 3(快速选举算法,基于 TCP),0 表示 Leader 选举算法(基于 UDP),1 表示非授权快速选举算法(基于 UDP),2 表示授权快速选举算法(基于 UDP),目前算法 1 和 2 都没有应用,所以不建议使用,算法 0 未来也可能会被淘汰,只保留算法 3,因此最好直接使用默认值。

(11) leaderServes

(Java system property:zookeeper.leaderServes)如果该值不是 no,则表示该服务器作为 Leader 时是需要接受客户端连接的。为了获得更高的吞吐量,当服务器数在 3 台以上时一般建议设置为 no。

(12) cnxTimeout

(Java system property:zookeeper.cnxTimeout)默认值是 5 000,单位为毫秒,表示 leaderelection 时打开连接的超时时间,只用在算法 3 中。

5. 将配置文件分发到集群其他机器上

```
[hadoop@hadoop1 apps]$ scp -r zookeeper-3.4.10/ hadoop2:$PWD
[hadoop@hadoop1 apps]$ scp -r zookeeper-3.4.10/ hadoop3:$PWD
```

然后在各个 ZooKeeper 服务器节点上新建目录 dataDir = /home/hadoop/data/zkdata,这个目录就是在 zoo.cfg 中配置的 dataDir 目录,建好之后在里面新建一个 myid 文件,其中存放的内容就是服务器的 ID,即 server.1 = hadoop01:2888:3888,对应的每个服务器节点都应该做类似的操作。

我们以服务器 hadoop1 为例,参考配置如下:

```
[hadoop@hadoop1 ~]$ mkdir /home/hadoop/data/zkdata
[hadoop@hadoop1 ~]$ cd data/zkdata/
[hadoop@hadoop1 zkdata]$ echo 1 > myid
```

当以上所有步骤都完成时,意味着 ZooKeeper 的配置文件相关的修改都做完了。

6. 配置环境变量

```
[hadoop@hadoop1 ~]$ vi .bashrc
#Zookeeper PATH
export ZOOKEEPER_HOME=/home/hadoop/apps/zookeeper-3.4.10
export PATH=$PATH:$ZOOKEEPER_HOME/bin
```

保存退出,然后使配置生效:

```
[hadoop@hadoop1 ~]$ source .bashrc
```

7. 启动软件,并验证安装是否成功

注意:虽然在配置文件中写明了服务器的列表信息,但是我们还是需要单独进行每一台服务器的启动,而不是一键启动集群模式。

启动：zkServer.sh start。

停止：zkServer.sh stop。

查看状态：zkServer.sh status。

每启动一台服务器，都要查看一下状态再启动下一台服务器。

(1) 启动 hadoop1

```
[hadoop@hadoop1 ~]$ zkServer.sh start
ZooKeeper JMX enabled by default
Using config: /home/hadoop/apps/zookeeper-3.4.10/bin/../conf/zoo.cfg
Starting zookeeper ... STARTED
[hadoop@hadoop1 ~]$ zkServer.sh status
ZooKeeper JMX enabled by default
Using config: /home/hadoop/apps/zookeeper-3.4.10/bin/../conf/zoo.cfg
Error contacting service. It is probably not running.
[hadoop@hadoop1 ~]$
```

(2) 启动 hadoop2

```
[hadoop@hadoop2 ~]$ zkServer.sh start
ZooKeeper JMX enabled by default
Using config: /home/hadoop/apps/zookeeper-3.4.10/bin/../conf/zoo.cfg
Starting zookeeper ... STARTED
[hadoop@hadoop2 ~]$ zkServer.sh status
ZooKeeper JMX enabled by default
Using config: /home/hadoop/apps/zookeeper-3.4.10/bin/../conf/zoo.cfg
Mode: leader
```

此时再查看 hadoop1 的状态：

```
ZooKeeper JMX enabled by default
Using config: /home/hadoop/apps/zookeeper-3.4.10/bin/../conf/zoo.cfg
Mode: follower
```

(3) 启动 hadoop3

```
[hadoop@hadoop3 ~]$ zkServer.sh start
ZooKeeper JMX enabled by default
Using config: /home/hadoop/apps/zookeeper-3.4.10/bin/../conf/zoo.cfg
Starting zookeeper ... STARTED
[hadoop@hadoop3 ~]$ zkServer.sh status
ZooKeeper JMX enabled by default
Using config: /home/Hadoop/apps/zookeeper-3.4.10/bin/../conf/zoo.cfg
Mode: follower
```

8. 查看进程

3 台机器上都有 QuorumPeerMain 进程，可以使用 jps 命令查看：

```
[hadoop@hadoop1 ~]$ jps
2499 Jps
2404 QuorumPeerMain
```

9．ZooKeeper 命令工具

在启动 ZooKeeper 服务之后，输入以下命令，连接到 ZooKeeper 服务：

[hadoop@hadoop1 ~]$ zkCli.sh -server hadoop2:2181
   Connecting to hadoop2:2181
   2020-05-27 16:15:53,744 [myid:] - INFO [main:Environment@100] - Client environment:zookeeper.version=3.4.10-39d3a4f269333c922ed3db283be479f9deacaa0f, built on 03/23/2017 10:13 GMT
   2020-05-27 16:15:53,748 [myid:] - INFO [main:Environment@100] - Client environment:host.name=Hadoop1
   2020-05-27 16:15:53,749 [myid:] - INFO [main:Environment@100] - Client environment:java.version=1.8.0_73
   2020-05-27 16:15:53,751 [myid:] - INFO [main:Environment@100] - Client environment:java.vendor=Oracle Corporation
   2020-05-27 16:15:53,751 [myid:] - INFO [main:Environment@100] - Client environment:java.home=/usr/local/jdk1.8.0_73/jre
   2020-05-27 16:15:53,751 [myid:] - INFO [main:Environment@100] - Client environment:java.class.path=/home/hadoop/apps/zookeeper-3.4.10/bin/../build/classes:/home/hadoop/apps/zookeeper-3.4.10/bin/../build/lib/*.jar:/home/hadoop/apps/zookeeper-3.4.10/bin/../lib/slf4j-log4j12-1.6.1.jar:/home/hadoop/apps/zookeeper-3.4.10/bin/../lib/slf4j-api-1.6.1.jar:/home/hadoop/apps/zookeeper-3.4.10/bin/../lib/netty-3.10.5.Final.jar:/home/Hadoop/apps/zookeeper-3.4.10/bin/../lib/log4j-1.2.16.jar:/home/hadoop/apps/zookeeper-3.4.10/bin/../lib/jline-0.9.94.jar:/home/hadoop/apps/zookeeper-3.4.10/bin/../zookeeper-3.4.10.jar:/home/hadoop/apps/zookeeper-3.4.10/bin/../src/java/lib/*.jar:/home/hadoop/apps/zookeeper-3.4.10/bin/../conf::/usr/local/jdk1.8.0_73/lib:/usr/local/jdk1.8.0_73/jre/lib
   2020-05-27 16:15:53,751 [myid:] - INFO [main:Environment@100] - Client environment:java.library.path=/usr/java/packages/lib/amd64:/usr/lib64:/lib64:/lib:/usr/lib
   2020-05-27 16:15:53,751 [myid:] - INFO [main:Environment@100] - Client environment:java.io.tmpdir=/tmp
   2020-05-27 16:15:53,751 [myid:] - INFO [main:Environment@100] - Client environment:java.compiler=<NA>
   2020-05-27 16:15:53,752 [myid:] - INFO [main:Environment@100] - Client environment:os.name=Linux
   2020-05-27 16:15:53,752 [myid:] - INFO [main:Environment@100] - Client environment:os.arch=amd64
   2020-05-27 16:15:53,752 [myid:] - INFO [main:Environment@100] - Client environment:os.version=2.6.32-573.el6.x86_64
   2020-05-27 16:15:53,752 [myid:] - INFO [main:Environment@100] - Client environment:user.name=Hadoop
   2020-05-27 16:15:53,752 [myid:] - INFO [main:Environment@100] - Client environment:user.home=/home/Hadoop
   2020-05-27 16:15:53,752 [myid:] - INFO [main:Environment@100] - Client environment:user.dir=/home/Hadoop
   2020-05-27 16:15:53,755 [myid:] - INFO [main:ZooKeeper@438] - Initiating client connection, connectString=hadoop2:2181 sessionTimeout=30000 watcher=org.apache.zookeeper.ZooKeeperMain$MyWatcher@5c29bfd
   Welcome to ZooKeeper!
   2020-05-27 16:15:53,789 [myid:] - INFO [main-SendThread(hadoop2:2181):ClientCnxn$SendThread@1032] - Opening socket connection to server hadoop2/192.168.123.103:2181. Will not attempt to authenticate using SASL (unknown error)
   JLine support is enabled

```
2020-05-27 16:15:53,931 [myid:] - INFO  [main-SendThread(Hadoop2:2181):ClientCnxn$SendThread@876] - Socket connection established to hadoop2/192.168.123.103:2181, initiating session
2020-05-27 16:15:53,977 [myid:] - INFO  [main-SendThread(hadoop2:2181):ClientCnxn$SendThread@1299] - Session establishment complete on server hadoop2/192.168.123.103:2181, sessionid = 0x262486284b70000, negotiated timeout = 30000
WATCHER::
WatchedEvent state:SyncConnected type:None path:null
[zk: hadoop2:2181(CONNECTED) 0]
```

连接成功之后，系统会输出 ZooKeeper 的相关环境及配置信息，并在屏幕上输出 "Welcome to Zookeeper!" 等信息。

8.4.2 使用 ZooKeeper 命令的简单操作步骤

① 使用 ls 命令查看当前 ZooKeeper 中所包含的内容：ls /。

```
[zk: hadoop2:2181(CONNECTED) 1] ls /
[zookeeper]
[zk: hadoop2:2181(CONNECTED) 2]
```

② 创建一个新的 Znode 节点 "aa"，以及和它相关的字符，执行命令：create /aa "my first zk"。默认是不带编号的。

```
[zk: hadoop2:2181(CONNECTED) 2] create /aa "my first zk"
Created /aa
[zk: hadoop2:2181(CONNECTED) 3]
```

创建带编号的永久节点 "bb"：

```
[zk: localhost:2181(CONNECTED) 1] create -s /bb "bb"
Created /bb0000000001
[zk: localhost:2181(CONNECTED) 2]
```

创建不带编号的临时节点 "cc"：

```
[zk: localhost:2181(CONNECTED) 2] create -e /cc "cc"
Created /cc
[zk: localhost:2181(CONNECTED) 3]
```

创建带编号的临时节点 "dd"：

```
[zk: localhost:2181(CONNECTED) 3] create -s -e /dd "dd"
Created /dd0000000003
[zk: localhost:2181(CONNECTED) 4]
```

③ 再次使用 ls 命令来查看现在 ZooKeeper 中所包含的内容：ls /。

```
[zk: localhost:2181(CONNECTED) 4] ls /
[cc, dd0000000003, zookeeper, bb0000000001]
[zk: localhost:2181(CONNECTED) 5]
```

可以看到，aa 节点已经被创建。关闭本次连接会话，再重新打开一个连接：

```
[zk: localhost:2181(CONNECTED) 5] close
2020-05-27 13:03:29,137 [myid:] - INFO [main:ZooKeeper@684] - Session: 0x1624c10e8d90000 closed
  [zk: localhost:2181(CLOSED) 6] 2020-05-27 13:03:29,139 [myid:] - INFO [main-EventThread:
ClientCnxn$EventThread@519] - EventThread shut down for session: 0x1624c10e8d90000
[zk: localhost:2181(CLOSED) 6] ls /
Not connected
[zk: localhost:2181(CLOSED) 7] connect hadoop1:2181
```

重新查看，临时节点已经随着上一次的会话关闭自动删除了：

```
[zk: hadoop1:2181(CONNECTED) 8] ls /
[zookeeper, bb0000000001]
[zk: hadoop1:2181(CONNECTED) 9]
```

④ 使用 get 命令来确认第②步中所创建的 Znode 是否包含我们创建的字符串，执行命令：get /aa。

```
[zk: hadoop2:2181(CONNECTED) 4] get /aa
my first zk
cZxid = 0x100000002
ctime = Wed Apr 27 20:01:02 CST 2020
mZxid = 0x100000002
mtime = Wed Apr 27 20:01:02 CST 2020
pZxid = 0x100000002
cversion = 0
dataVersion = 0
aclVersion = 0
ephemeralOwner = 0x0
dataLength = 11
numChildren = 0
[zk: hadoop2:2181(CONNECTED) 5]
```

⑤ 接下来通过 set 命令对 Znode 所关联的字符串进行设置，执行命令：set /aa mytest。

```
[zk: hadoop2:2181(CONNECTED) 6] set /aa mytest
cZxid = 0x100000002
ctime = Wed Apr 27 20:01:02 CST 2020
mZxid = 0x100000004
mtime = Wed Apr 27 20:04:10 CST 2020
pZxid = 0x100000002
cversion = 0
dataVersion = 1
aclVersion = 0
ephemeralOwner = 0x0
dataLength = 7
numChildren = 0
[zk: hadoop2:2181(CONNECTED) 7]
```

⑥ 再次使用 get 命令查看上次修改的内容，执行命令：get /aa。

```
[zk:hadoop2:2181(CONNECTED) 7] get /aa
mytest
cZxid = 0x100000002
ctime = Wed Apr 27 20:01:02 CST 2020
mZxid = 0x100000004
mtime = Wed Apr 27 20:04:10 CST 2020
pZxid = 0x100000002
cversion = 0
dataVersion = 1
aclVersion = 0
ephemeralOwner = 0x0
dataLength = 7
numChildren = 0
[zk:hadoop2:2181(CONNECTED) 8]
```

⑦ 将刚才创建的 Znode 删除，执行命令：delete /aa。

```
[zk:hadoop2:2181(CONNECTED) 8] delete /aa
[zk:hadoop2:2181(CONNECTED) 9]
```

⑧ 再次使用 ls 命令查看 ZooKeeper 中的内容，执行命令：ls /。

```
[zk:hadoop2:2181(CONNECTED) 9] ls /
[zookeeper]
[zk:hadoop2:2181(CONNECTED) 10]
```

经过验证，Znode 节点已被删除。

⑨ 退出，执行命令：quit。

```
[zk:hadoop2:2181(CONNECTED) 10] quit
Quitting...
2020-05-27 20:07:11,133 [myid:] - INFO  [main:ZooKeeper@684] - Session: 0x262486284b70000 closed
2020-05-27 20:07:11,139 [myid:] - INFO  [main-EventThread:ClientCnxn $ EventThread @ 519] - EventThread shut down for session: 0x262486284b70000
[hadoop@hadoop1 ~]$
```

本 章 小 结

① ZooKeeper 是 Apache 旗下的一个软件项目，它为大型分布式计算提供开源的分布式配置服务、同步服务和命名注册。

② ZooKeeper 的架构通过冗余服务实现高可用性。ZooKeeper 的设计目标是将那些复杂且容易出错的分布式一致性服务封装起来，构成一个高效可靠的原语集，并以一系列简单易用的接口提供给用户使用。

③ ZooKeeper 是一个典型的分布式数据一致性的解决方案,分布式应用程序可以基于它实现数据发布/订阅、负载均衡、命名服务、分布式协调/通知、集群管理、Master 选举、分布式锁和分布式队列等功能。

第 9 章
Sqoop 数据迁移工具

9.1 Sqoop 功能概述

9.1.1 Sqoop 软件介绍

Sqoop 是 Apache 旗下的一款 Hadoop 和关系数据库服务器之间传送数据的工具。Sqoop 是一款开源的工具,主要用于在 HDFS 与传统的数据库(如 MySQL、Oracle、PostgreSQL 等)之间进行数据的传递,可以将一个关系数据库(如 MySQL、Oracle、PostgreSQL 等)中的数据导入 Hadoop 的 HDFS 中,也可以将 HDFS 中的数据导入关系数据库中。

1. 核心功能

Sqoop 的核心功能有两个:导入和导出数据。

导入数据:从 MySQL、Oracle 中导入数据到 Hadoop 的 HDFS、Hive、HBase 等数据存储系统。

导出数据:从 Hadoop 的文件系统中导出数据到关系数据库(如 MySQL 等)。

Sqoop 的本质还是一个命令行工具,和 HDFS、Hive 相比,并没有什么高深的理论,主要工作原理如图 9-1 所示。

- Sqoop 工具:本质就是迁移数据,迁移的方式是把 Sqoop 的迁移命令转换成 MapReduce 程序。
- Hive 工具:本质就是执行计算,依赖于 HDFS 存储数据,把 SQL 转换成 MapReduce 程序。

2. 工作机制

将导入或导出命令翻译成 MapReduce 程序来实现。在翻译出的 MapReduce 中主要是对 InputFormat 和 OutputFormat 进行定制。

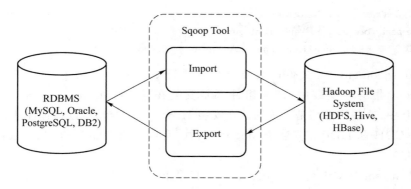

图 9-1　Sqoop 主要工作原理

9.1.2　Sqoop 软件安装

1. 软件下载

可以从 Apache 官方网站 http://apache.hadoop.org 下载 Sqoop 软件。但建议从国内镜像站点下载，参考地址为 http://mirrors.hust.edu.cn/apache/。

目前绝大部分企业所使用的 Sqoop 的版本都是 Sqoop 版本 1，Sqoop 1.4.6 或者 Sqoop 1.4.7 指的是 Sqoop 版本 1，Sqoop 1.99.4 指的是 Sqoop 版本 2。本书使用 Sqoop 1.4.6 版本 sqoop-1.4.6.bin__hadoop-2.0.4-alpha.tar.gz。

2. 安装步骤

（1）上传 Sqoop 软件，解压安装包到指定目录

因为之前 Hive 只是安装在 hadoop3 主机上，所以 Sqoop 同样安装在 hadoop3 主机上。

```
[hadoop@hadoop3 ~]$ tar -zxvf sqoop-1.4.6.bin__hadoop-2.0.4-alpha.tar.gz -C apps/
```

（2）进入 conf 文件夹，找到 sqoop-env-template.sh，修改其名称为 sqoop-env.sh

```
[hadoop@hadoop3 ~]$ cd apps/
[hadoop@hadoop3 apps]$ mv sqoop-1.4.6.bin__hadoop-2.0.4-alpha/ sqoop-1.4.6
[hadoop@hadoop3 apps]$ cd sqoop-1.4.6/conf/
[hadoop@hadoop3 conf]$ ls
oraoop-site-template.xml  sqoop-env-template.sh  sqoop-site.xml
sqoop-env-template.cmd    sqoop-site-template.xml
[hadoop@hadoop3 conf]$ mv sqoop-env-template.sh sqoop-env.sh
```

（3）修改 sqoop-env.sh

```
[hadoop@hadoop3 conf]$ vi sqoop-env.sh
export HADOOP_COMMON_HOME=/home/hadoop/apps/hadoop-2.7.5
#Set path to wherehadoop-*-core.jar is available
export HADOOP_MAPRED_HOME=/home/hadoop/apps/hadoop-2.7.5
#Set the path to where bin/hbase is available
export HBASE_HOME=/home/hadoop/apps/hbase-1.2.6
#Set the path to where bin/hive is available
```

```
export HIVE_HOME = /home/hadoop/apps/apache-hive-2.3.8-bin
#Set the path for where zookeper config dir is
export ZOOCFGDIR = /home/hadoop/apps/zookeeper-3.4.10/conf
```

为什么在 sqoop-env.sh 文件中会要求分别进行 common 和 MapReduce 的配置呢？在 Apache Hadoop 集群的安装中，四大组件都是安装在同一个 hadoop_home 中的。但是在 CDH 和 HDP 中，这些组件都是可选的。在安装 Hadoop 的时候，可以选择性地只安装 HDFS 或者 YARN，CDH 和 HDP 在安装 Hadoop 的时候，有可能把 HDFS 和 MapReduce 分别安装在不同的地方。

（4）加入 mysql 驱动包到 sqoop-1.4.6/lib 目录下

[hadoop@hadoop3 ~]$ cp mysql-connector-java-5.1.40-bin.jar apps/sqoop-1.4.6/lib/

（5）配置系统环境变量

[hadoop@hadoop3 ~]$ vim .bashrc

```
#Sqoop configure
export SQOOP_HOME = /home/hadoop/apps/sqoop-1.4.6
export PATH = $PATH:$SQOOP_HOME/bin
```

保存退出，使其立即生效：

[hadoop@hadoop3 ~]$ source .bashrc

（6）验证安装是否成功

[hadoop@hadoop3 ~]$ sqoop-version 或者 sqoop version

9.2 Sqoop 命令操作

9.2.1 Sqoop 的基本命令

1. help 操作

首先，我们可以使用 sqoop help 来查看 Sqoop 支持哪些命令：

```
[hadoop@hadoop3 ~]$ sqoop help
Warning: /home/Hadoop/apps/sqoop-1.4.6/../hcatalog does not exist! HCatalog jobs will fail.
Please set $HCAT_HOME to the root of your HCatalog installation.
Warning: /home/hadoop/apps/sqoop-1.4.6/../accumulo does not exist! Accumulo imports will fail.
Please set $ACCUMULO_HOME to the root of your Accumulo installation.
20/05/12 13:37:19 INFO sqoop.Sqoop: Running Sqoop version: 1.4.6
usage: sqoop COMMAND [ARGS]
Available commands:
  codegen              Generate code to interact with database records
  create-hive-table    Import a table definition into Hive
  eval                 Evaluate a SQL statement and display the results
```

```
  export                      Export an HDFS directory to a database table
  help                        List available commands
  import                      Import a table from a database to HDFS
  import-all-tables           Import tables from a database to HDFS
  import-mainframe            Import datasets from a mainframe server to HDFS
  job                         Work with saved jobs
  list-databases              List available databases on a server
  list-tables                 List available tables in a database
  merge                       Merge results of incremental imports
  metastore                   Run a standalone Sqoop metastore
  version                     Display version information
See 'sqoop help COMMAND' for information on a specific command.
[hadoop@hadoop3 ~]$
```

得到这些支持的命令之后，如果不知道使用方式，可以使用 sqoop command 命令来查看某条具体命令的使用方式，例如：

```
[hadoop@hadoop3 ~]$ sqoop help import
Warning: /home/hadoop/apps/sqoop-1.4.6/../hcatalog does not exist! HCatalog jobs will fail.
Please set $HCAT_HOME to the root of your HCatalog installation.
Warning: /home/hadoop/apps/sqoop-1.4.6/../accumulo does not exist! Accumulo imports will fail.
Please set $ACCUMULO_HOME to the root of your Accumulo installation.
20/05/12 13:38:29 INFO sqoop.Sqoop: Running Sqoop version: 1.4.6
usage: sqoop import [GENERIC-ARGS] [TOOL-ARGS]
Common arguments:
   --connect <jdbc-uri>                         Specify JDBC connect
```

2. 列出 MySQL 有哪些数据库

```
[hadoop@hadoop3 ~]$ sqoop list-databases \
> --connect jdbc:mysql://hadoop1:3306/  --username root --password root
Warning: /home/hadoop/apps/sqoop-1.4.6/../hcatalog does not exist! HCatalog jobs will fail.
Please set $HCAT_HOME to the root of your HCatalog installation.
Warning: /home/hadoop/apps/sqoop-1.4.6/../accumulo does not exist! Accumulo imports will fail.
Please set $ACCUMULO_HOME to the root of your Accumulo installation.
20/05/12 13:43:51 INFO sqoop.Sqoop: Running Sqoop version: 1.4.6
20/05/12 13:43:51 WARN tool.BaseSqoopTool: Setting your password on the command-line is insecure. Consider using -P instead.
20/05/12 13:43:51 INFO manager.MySQLManager: Preparing to use a MySQL streaming resultset.
information_schema
hivedb
mysql
performance_schema
test
[hadoop@hadoop3 ~]$
```

3. 列出 MySQL 中的某个数据库有哪些数据表

```
[hadoop@hadoop3 ~]$ sqoop list-tables \
> --connect jdbc:mysql://Hadoop1:3306/mysql --username root --password root
Warning: /home/hadoop/apps/sqoop-1.4.6/../hcatalog does not exist! HCatalog jobs will fail.
Please set $HCAT_HOME to the root of your HCatalog installation.
Warning: /home/hadoop/apps/sqoop-1.4.6/../accumulo does not exist! Accumulo imports will fail.
Please set $ACCUMULO_HOME to the root of your Accumulo installation.
20/05/12 13:46:21 INFO sqoop.Sqoop: Running Sqoop version: 1.4.6
20/05/12 13:46:21 WARN tool.BaseSqoopTool: Setting your password on the command-line is insecure. Consider using -P instead.
20/05/12 13:46:21 INFO manager.MySQLManager: Preparing to use a MySQL streaming resultset.
columns_priv
db
event
func
general_log
help_category
help_keyword
help_relation
help_topic
innodb_index_stats
innodb_table_stats
ndb_binlog_index
plugin
proc
procs_priv
proxies_priv
servers
slave_master_info
slave_relay_log_info
slave_worker_info
slow_log
tables_priv
time_zone
time_zone_leap_second
time_zone_name
time_zone_transition
time_zone_transition_type
user
[hadoop@hadoop3 ~]$
```

4. 以复制方式创建表

创建一张和 MySQL 中的 help_keyword 表一样的 Hive 表 help_keyword2：

```
[hadoop@hadoop3 ~]$ sqoop create-hive-table \
--connect jdbc:mysql://hadoop1:3306/mysql --username root --password root \
```

    --table help_keyword --hive-table help_keyword2

> Warning: /home/hadoop/apps/sqoop-1.4.6/../hcatalog does not exist! HCatalog jobs will fail.
> Please set $HCAT_HOME to the root of your HCatalog installation.
> Warning: /home/hadoop/apps/sqoop-1.4.6/../accumulo does not exist! Accumulo imports will fail.
> Please set $ACCUMULO_HOME to the root of your Accumulo installation.
> 20/05/12 13:50:20 INFO sqoop.Sqoop: Running Sqoop version: 1.4.6
> 20/05/12 13:50:20 WARN tool.BaseSqoopTool: Setting your password on the command-line is insecure. Consider using -P instead.
> 20/05/12 13:50:20 INFO tool.BaseSqoopTool: Using Hive-specific delimiters for output. You can override
> 20/05/12 13:50:20 INFO tool.BaseSqoopTool: delimiters with --fields-terminated-by, etc.
> 20/05/12 13:50:20 INFO manager.MySQLManager: Preparing to use a MySQL streaming resultset.
> 20/05/12 13:50:21 INFO manager.SqlManager: Executing SQL statement: SELECT t.* FROM `help_keyword` AS t LIMIT 1
> 20/05/12 13:50:21 INFO manager.SqlManager: Executing SQL statement: SELECT t.* FROM `help_keyword` AS t LIMIT 1
> SLF4J: Class path contains multiple SLF4J bindings.
> SLF4J: Found binding in [jar:file:/home/hadoop/apps/hadoop-2.7.5/share/hadoop/common/lib/slf4j-log4j12-1.7.10.jar!/org/slf4j/impl/StaticLoggerBinder.class]
> SLF4J: Found binding in [jar:file:/home/hadoop/apps/hbase-1.2.6/lib/slf4j-log4j12-1.7.5.jar!/org/slf4j/impl/StaticLoggerBinder.class]
> SLF4J: See http://www.slf4j.org/codes.html#multiple_bindings for an explanation.
> SLF4J: Actual binding is of type [org.slf4j.impl.Log4jLoggerFactory]
> 20/05/12 13:50:23 INFO hive.HiveImport: Loading uploaded data into Hive
> 20/05/12 13:50:34 INFO hive.HiveImport: SLF4J: Class path contains multiple SLF4J bindings.
> 20/05/12 13:50:34 INFO hive.HiveImport: SLF4J: Found binding in [jar:file:/home/hadoop/apps/apache-hive-2.3.8-bin/lib/log4j-slf4j-impl-2.6.2.jar!/org/slf4j/impl/StaticLoggerBinder.class]
> 20/05/12 13:50:34 INFO hive.HiveImport: SLF4J: Found binding in [jar:file:/home/hadoop/apps/hbase-1.2.6/lib/slf4j-log4j12-1.7.5.jar!/org/slf4j/impl/StaticLoggerBinder.class]
> 20/05/12 13:50:34 INFO hive.HiveImport: SLF4J: Found binding in [jar:file:/home/hadoop/apps/Hadoop-2.7.5/share/hadoop/common/lib/slf4j-log4j12-1.7.10.jar!/org/slf4j/impl/StaticLoggerBinder.class]
> 20/05/12 13:50:34 INFO hive.HiveImport: SLF4J: See http://www.slf4j.org/codes.html#multiple_bindings for an explanation.
> 20/05/12 13:50:34 INFO hive.HiveImport: SLF4J: Actual binding is of type [org.apache.logging.slf4j.Log4jLoggerFactory]
> 20/05/12 13:50:36 INFO hive.HiveImport:
> 20/05/12 13:50:36 INFO hive.HiveImport: Logging initialized using configuration in jar:file:/home/hadoop/apps/apache-hive-2.3.8-bin/lib/hive-common-2.3.3.jar!/hive-log4j2.properties Async: true
> 20/05/12 13:50:50 INFO hive.HiveImport: OK
> 20/05/12 13:50:50 INFO hive.HiveImport: Time taken: 11.651 seconds
> 20/05/12 13:50:51 INFO hive.HiveImport: Hive import complete.

9.2.2 Sqoop 的数据导入

导入工具从 RDBMS 导入单个表到 HDFS，表中的每一行被视为 HDFS 的记录。所有记录都存储为文本文件的文本数据（或者 Avro、Sequence 文件等二进制数据）。

1. 从 RDBMS 导入 HDFS 中

语法格式：sqoop import（generic-args 参数）（import-args 参数）。

常用参数如下。

--connect < jdbc-uri >：JDBC 连接地址。

--connection-manager < class-name >：连接管理者。

--driver < class-name >：驱动类。

--hadoop-mapred-home < dir >：$ HADOOP_MAPRED_HOME。

--help：help 信息。

-P：从命令行输入密码。

--password < password >：密码。

--username < username >：账号。

--verbose：打印流程信息。

--connection-param-file < filename >：可选参数。

普通导入：导入 MySQL 库中的 help_keyword 的数据到 HDFS 上，导入的默认路径为 /user/hadoop/help_keyword。

```
[hadoop@hadoop3 ~]$ sqoop import  \
--connect jdbc:mysql://hadoop1:3306/mysql  --username root  --password root  \
--table help_keyword  -m 1
Warning: /home/hadoop/apps/sqoop-1.4.6/../hcatalog does not exist! HCatalog jobs will fail.
Please set $HCAT_HOME to the root of your HCatalog installation.
Warning: /home/hadoop/apps/sqoop-1.4.6/../accumulo does not exist! Accumulo imports will fail.
Please set $ACCUMULO_HOME to the root of your Accumulo installation.
20/05/12 13:53:48 INFO sqoop.Sqoop: Running Sqoop version: 1.4.6
20/05/12 13:53:48 WARN tool.BaseSqoopTool: Setting your password on the command-line is insecure. Consider using -P instead.
20/05/12 13:53:48 INFO manager.MySQLManager: Preparing to use a MySQL streaming resultset.
20/05/12 13:53:48 INFO tool.CodeGenTool: Beginning code generation
20/05/12 13:53:49 INFO manager.SqlManager: Executing SQL statement: SELECT t.* FROM `help_keyword` AS t LIMIT 1
20/05/12 13:53:49 INFO manager.SqlManager: Executing SQL statement: SELECT t.* FROM `help_keyword` AS t LIMIT 1
20/05/12 13:53:49 INFO orm.CompilationManager: HADOOP_MAPRED_HOME is /home/hadoop/apps/hadoop-2.7.5
#/tmp/sqoop-hadoop/compile/979d87b9521d0a09ee6620060a112d60/help_keyword.java
20/05/12 13:53:51 INFO orm.CompilationManager: Writing jar file: /tmp/sqoop-Hadoop/compile/979d87b9521d0a09ee6620060a112d60/help_keyword.jar
20/05/12 13:53:51 WARN manager.MySQLManager: It looks like you are importing from mysql.
20/05/12 13:53:51 WARN manager.MySQLManager: This transfer can be faster! Use the --direct
20/05/12 13:53:51 WARN manager.MySQLManager: option to exercise a MySQL-specific fast path.
20/05/12 13:53:51 INFO manager.MySQLManager: Setting zero DATETIME behavior to convertToNull (mysql)
20/05/12 13:53:51 INFO MapReduce.ImportJobBase: Beginning import of help_keyword
SLF4J: Class path contains multiple SLF4J bindings.
```

SLF4J: Found binding in [jar:file:/home/hadoop/apps/hadoop-2.7.5/share/hadoop/common/lib/slf4j-log4j12-1.7.10.jar!/org/slf4j/impl/StaticLoggerBinder.class]
SLF4J: Found binding in [jar:file:/home/hadoop/apps/hbase-1.2.6/lib/slf4j-log4j12-1.7.5.jar!/org/slf4j/impl/StaticLoggerBinder.class]
SLF4J: See http://www.slf4j.org/codes.html#multiple_bindings for an explanation.
SLF4J: Actual binding is of type [org.slf4j.impl.Log4jLoggerFactory]
20/05/12 13:53:52 INFO Configuration.deprecation: mapred.jar is deprecated. Instead, use MapReduce.job.jar
20/05/12 13:53:53 INFO Configuration.deprecation: mapred.map.tasks is deprecated. Instead, use MapReduce.job.maps
20/05/12 13:53:58 INFO db.DBInputFormat: Using read commited transaction isolation
20/05/12 13:53:58 INFO MapReduce.JobSubmitter: number of splits:1
20/05/12 13:53:59 INFO MapReduce.JobSubmitter: Submitting tokens for job: job_1523510178850_0001
20/05/12 13:54:00 INFO impl.YarnClientImpl: Submitted application application_1523510178850_0001
20/05/12 13:54:00 INFO MapReduce.Job: The url to track the job: http://hadoop3:8088/proxy/application_1523510178850_0001/
20/05/12 13:54:00 INFO MapReduce.Job: Running job: job_1523510178850_0001
20/05/12 13:54:17 INFO MapReduce.Job: Job job_1523510178850_0001 running in uber mode : false
20/05/12 13:54:17 INFO MapReduce.Job: map 0% reduce 0%
20/05/12 13:54:33 INFO MapReduce.Job: map 100% reduce 0%
20/05/12 13:54:34 INFO MapReduce.Job: Job job_1523510178850_0001 completed successfully
20/05/12 13:54:35 INFO MapReduce.Job: Counters: 30
    File System Counters
        FILE: Number of bytes read = 0
        FILE: Number of bytes written = 142965
        FILE: Number of read operations = 0
        FILE: Number of large read operations = 0
        FILE: Number of write operations = 0
        HDFS: Number of bytes read = 87
        HDFS: Number of bytes written = 8264
        HDFS: Number of read operations = 4
        HDFS: Number of large read operations = 0
        HDFS: Number of write operations = 2
    Job Counters
        Launched map tasks = 1
        Other local map tasks = 1
        Total time spent by all maps in occupied slots (ms) = 12142
        Total time spent by all reduces in occupied slots (ms) = 0
        Total time spent by all map tasks (ms) = 12142
        Total vcore-milliseconds taken by all map tasks = 12142
        Total megabyte-milliseconds taken by all map tasks = 12433408
    Map-Reduce Framework
        Map input records = 619
        Map output records = 619
        Input split bytes = 87
        Spilled Records = 0

```
                Failed Shuffles = 0
                Merged Map outputs = 0
                GC time elapsed (ms) = 123
                CPU time spent (ms) = 1310
                Physical memory (bytes) snapshot = 93212672
                Virtual memory (bytes) snapshot = 2068234240
                Total committed heap usage (bytes) = 17567744
        File Input Format Counters
                Bytes Read = 0
        File Output Format Counters
                Bytes Written = 8264
20/05/12 13:54:35 INFO MapReduce.ImportJobBase: Transferred 8.0703 KB in 41.8111 seconds
(197.6507 bytes/sec)
20/05/12 13:54:35 INFO MapReduce.ImportJobBase: Retrieved 619 records.
[hadoop@hadoop3 ~]$
```

查看导入的文件：

```
[hadoop@hadoop3 ~]$ hadoop fs -cat /user/hadoop/help_keyword/part-m-00000
    0,JOIN
    1,HOST
    2,SERIALIZABLE
    3,CONTAINS
    4,SRID
    5,AT
    6,SCHEDULE
    7,RETURNS
    8,MASTER_SSL_CA
    9,NCHAR
    10,ONLY
    11,ST_GEOMETRYN
    12,ST_GEOMCOLLFROMWKB
    13,WORK
    14,OPEN
    15,SET_INTERSECTS
    16,ESCAPE
    17,EVENTS
```

2. 指定分隔符和导入路径

```
[hadoop@hadoop3 ~]$ sqoop import \
--connect jdbc:mysql://hadoop1:3306/mysql  --username root --password root \
--table help_keyword \
--target-dir /user/hadoop11/my_help_keyword1 \
--fields-terminated-by '\t'  -m 2
```

3. 带 where 条件导入数据

```
[hadoop@hadoop3 ~]$ sqoop import \
```

```
--connect jdbc:mysql://hadoop1:3306/mysql   --username root   --password root   \
--where "name='STRING'"  \
--table help_keyword   \
--target-dir /sqoop/hadoop11/myoutput1   -m 1
```

4. 查询指定列

```
[hadoop@hadoop3 ~]$ sqoop import   \
--connect jdbc:mysql://hadoop1:3306/mysql   \
--username root   --password root   \
--columns "name" \
--where "name='STRING'"  \
--table help_keyword   \
--target-dir /sqoop/hadoop11/myoutput22   -m 1
```

验证数据：

```
hive>  select name from help_keyword where name = "string"
```

5. 指定自定义查询 SQL 导入

```
[hadoop@hadoop3 ~]$ sqoop import   \
--connect jdbc:mysql://hadoop1:3306/   \
--username root   --password root   \
--target-dir /user/hadoop/myimport33_1   \
--query 'select help_keyword_id,name from mysql.help_keyword where $CONDITIONS and name = "STRING"' \
--split-by  help_keyword_id \
--fields-terminated-by '\t'  \
-m 4
```

在以上需要按照自定义 SQL 语句导出数据到 HDFS 的情况下：
- 引号问题：要么外层使用单引号，内层使用双引号，$CONDITIONS 的 $ 符号不用转义；要么外层使用双引号，内层使用单引号，$CONDITIONS 的 $ 符号需要转义。
- 自定义的 SQL 语句中必须带有 WHERE \$CONDITIONS。

9.2.3 将 MySQL 数据库中的表数据导入 Hive

1. 将 MySQL 数据库中的表数据导入 Hive

Sqoop 导入关系型数据到 Hive 的过程是先将数据导入 HDFS，然后再 load 进入 Hive。普通导入：数据存储在默认的 Default 库中，表名就是对应的 MySQL 的表名。

```
[hadoop@hadoop3 ~]$ sqoop import   \
--connect jdbc:mysql://hadoop1:3306/mysql   \
--username root   --password root   \
--table help_keyword   \
--hive-import -m 1
```

导入过程可以分解为如下步骤。

第一步：导入 mysql.help_keyword 的数据到 HDFS 的默认路径。

第二步：自动复制 mysql.help_keyword 去创建一张 Hive 表，创建在默认的 Default 库中。

第三步：把临时目录中的数据导入 Hive 表中。

查看数据，可以输入如下命令：

```
[hadoop@hadoop3 ~]$ hadoop fs -cat /user/hive/warehouse/help_keyword/part-m-00000
```

2. 指定参数将 MySQL 数据库中的表数据导入 Hive

可以通过参数来指定 hive-import、覆盖导入、自动创建 Hive 表、表名、删除中间结果数据目录：

```
[hadoop@hadoop3 ~]$ sqoop import \
--connect jdbc:mysql://hadoop1:3306/mysql \
--username root  --password root \
--table help_keyword \
--fields-terminated-by "\t" \
--lines-terminated-by "\n" \
--hive-import \
--hive-overwrite \
--create-hive-table \
--delete-target-dir \
--hive-database  mydb_test \
--hive-table new_help_keyword
```

主要需要提前在 Hive 中创建 mydb_test 数据库，不然会出现目标数据库不存在的错误。报错原因是 hive-import 这个导入命令。Sqoop 会自动创建 Hive 表，但是不会自动创建不存在的库，需要手动创建 mydb_test 数据库。

```
hive> create database mydb_test;
OK
Time taken: 6.147 seconds
hive>
```

可以通过查询来验证 Hive 数据：

```
hive> use mydb_test;
hive> select * from new_help_keyword limit 10;
OK
0    JOIN
1    HOST
2    SERIALIZABLE
3    CONTAINS
4    SRID
5    AT
6    SCHEDULE
7    RETURNS
8    MASTER_SSL_CA
9    NCHAR
```

我们可以尝试用不同的参数来实现，上面的导入语句等价于：

```
[hadoop@hadoop3 ~]$ sqoop import \
--connect jdbc:mysql://hadoop1:3306/mysql \
--username root  --password root \
--table help_keyword \
--fields-terminated-by "\t" \
--lines-terminated-by "\n" \
--hive-import \
--hive-overwrite \
--create-hive-table \
--hive-table  mydb_test.new_help_keyword \
--delete-target-dir
```

3. 增量导入

执行增量导入之前，先清空 Hive 数据库中的 help_keyword 表的数据：

```
hive> truncate table help_keyword;
[hadoop@hadoop3 ~]$ sqoop import \
--connect jdbc:mysql://hadoop1:3306/mysql \
--username root  --password root \
--table help_keyword \
--target-dir /user/hadoop/myimport_add \
--incremental  append \
--check-column  help_keyword_id \
--last-value 500  -m 1
```

语句执行成功：

```
[hadoop@hadoop3 ~]$ sqoop import \
> --connect jdbc:mysql://hadoop1:3306/mysql \
> --username root \
> --password root \
> --table help_keyword \
> --target-dir /user/Hadoop/myimport_add \
> --incremental  append \
> --check-column  help_keyword_id \
> --last-value 500 \
> -m 1
Warning: /home/hadoop/apps/sqoop-1.4.6/../hcatalog does not exist! HCatalog jobs will fail.
Please set $HCAT_HOME to the root of your HCatalog installation.
Warning: /home/hadoop/apps/sqoop-1.4.6/../accumulo does not exist! Accumulo imports will fail.
Please set $ACCUMULO_HOME to the root of your Accumulo installation.
20/05/13 10:04:07 INFO sqoop.Sqoop: Running Sqoop version: 1.4.6
20/05/13 10:04:08 WARN tool.BaseSqoopTool: Setting your password on the command-line is insecure. Consider using -P instead.
```

20/05/13 10:04:08 INFO manager.MySQLManager: Preparing to use a MySQL streaming resultset.
20/05/13 10:04:08 INFO tool.CodeGenTool: Beginning code generation
20/05/13 10:04:08 INFO manager.SqlManager: Executing SQL statement: SELECT t.* FROM `help_keyword` AS t LIMIT 1
20/05/13 10:04:08 INFO manager.SqlManager: Executing SQL statement: SELECT t.* FROM `help_keyword` AS t LIMIT 1
20/05/13 10:04:08 INFO orm.CompilationManager: HADOOP_MAPRED_HOME is /home/hadoop/apps/hadoop-2.7.5
20/05/13 10:04:11 INFO orm.CompilationManager: Writing jar file: /tmp/sqoop-hadoop/compile/a51619d1ef8c6e4b112a209326ed9e0f/help_keyword.jar
SLF4J: Class path contains multiple SLF4J bindings.
SLF4J: Found binding in [jar:file:/home/Hadoop/apps/Hadoop-2.7.5/share/hadoop/common/lib/slf4j-log4j12-1.7.10.jar!/org/slf4j/impl/StaticLoggerBinder.class]
SLF4J: Found binding in [jar:file:/home/Hadoop/apps/hbase-1.2.6/lib/slf4j-log4j12-1.7.5.jar!/org/slf4j/impl/StaticLoggerBinder.class]
SLF4J: See http://www.slf4j.org/codes.html#multiple_bindings for an explanation.
SLF4J: Actual binding is of type [org.slf4j.impl.Log4jLoggerFactory]
20/05/13 10:04:12 INFO tool.ImportTool: Maximal id query for free form incremental import: SELECT MAX(`help_keyword_id`) FROM `help_keyword`
20/05/13 10:04:12 INFO tool.ImportTool: Incremental import based on column `help_keyword_id`
20/05/13 10:04:12 INFO tool.ImportTool: Lower bound value: 500
20/05/13 10:04:12 INFO tool.ImportTool: Upper bound value: 618
20/05/13 10:04:12 WARN manager.MySQLManager: It looks like you are importing from mysql.
20/05/13 10:04:12 WARN manager.MySQLManager: This transfer can be faster! Use the --direct
20/05/13 10:04:12 WARN manager.MySQLManager: option to exercise a MySQL-specific fast path.
20/05/13 10:04:12 INFO manager.MySQLManager: Setting zero DATETIME behavior to convertToNull (mysql)
20/05/13 10:04:12 INFO MapReduce.ImportJobBase: Beginning import of help_keyword
20/05/13 10:04:12 INFO Configuration.deprecation: mapred.jar is deprecated. Instead, use MapReduce.job.jar
20/05/13 10:04:12 INFO Configuration.deprecation: mapred.map.tasks is deprecated. Instead, use MapReduce.job.maps
20/05/13 10:04:17 INFO db.DBInputFormat: Using read commited transaction isolation
20/05/13 10:04:17 INFO MapReduce.JobSubmitter: number of splits:1
20/05/13 10:04:17 INFO MapReduce.JobSubmitter: Submitting tokens for job: job_1523510178850_0010
20/05/13 10:04:19 INFO impl.YarnClientImpl: Submitted application application_1523510178850_0010
20/05/13 10:04:19 INFO MapReduce.Job: The url to track the job: http://hadoop3:8088/proxy/application_1523510178850_0010/
20/05/13 10:04:19 INFO MapReduce.Job: Running job: job_1523510178850_0010
20/05/13 10:04:30 INFO MapReduce.Job: Job job_1523510178850_0010 running in uber mode : false
20/05/13 10:04:30 INFO MapReduce.Job: map 0% reduce 0%
20/05/13 10:04:40 INFO MapReduce.Job: map 100% reduce 0%
20/05/13 10:04:40 INFO MapReduce.Job: Job job_1523510178850_0010 completed successfully
20/05/13 10:04:41 INFO MapReduce.Job: Counters: 30
    File System Counters
        FILE: Number of bytes read=0

```
        FILE: Number of bytes written = 143200
        FILE: Number of read operations = 0
        FILE: Number of large read operations = 0
        FILE: Number of write operations = 0
        HDFS: Number of bytes read = 87
        HDFS: Number of bytes written = 1576
        HDFS: Number of read operations = 4
        HDFS: Number of large read operations = 0
        HDFS: Number of write operations = 2
    Job Counters
        Launched map tasks = 1
        Other local map tasks = 1
        Total time spent by all maps in occupied slots (ms) = 7188
        Total time spent by all reduces in occupied slots (ms) = 0
        Total time spent by all map tasks (ms) = 7188
        Total vcore-milliseconds taken by all map tasks = 7188
        Total megabyte-milliseconds taken by all map tasks = 7360512
    Map-Reduce Framework
        Map input records = 118
        Map output records = 118
        Input split bytes = 87
        Spilled Records = 0
        Failed Shuffles = 0
        Merged Map outputs = 0
        GC time elapsed (ms) = 86
        CPU time spent (ms) = 870
        Physical memory (bytes) snapshot = 95576064
        Virtual memory (bytes) snapshot = 2068234240
        Total committed heap usage (bytes) = 18608128
    File Input Format Counters
        Bytes Read = 0
    File Output Format Counters
        Bytes Written = 1576
20/05/13 10:04:41 INFO MapReduce.ImportJobBase: Transferred 1.5391 KB in 28.3008 seconds (55.6875 bytes/sec)
20/05/13 10:04:41 INFO MapReduce.ImportJobBase: Retrieved 118 records.
20/05/13 10:04:41 INFO util.AppendUtils: Creating missing output directory - myimport_add
20/05/13 10:04:41 INFO tool.ImportTool: Incremental import complete! To run another incremental import of all data following this import, supply the following arguments:
20/05/13 10:04:41 INFO tool.ImportTool:  --incremental append
20/05/13 10:04:41 INFO tool.ImportTool:  --check-column help_keyword_id
20/05/13 10:04:41 INFO tool.ImportTool:  --last-value 618
20/05/13 10:04:41 INFO tool.ImportTool: (Consider saving this with 'sqoop job --create')
```

查看结果：

```
[hadoop@hadoop3 ~]$ hadoop fs -cat /usr/hadoop/myimport_add/part-m-0000
```

部分输出结果如下:

```
501,HEAP
502,RETURNED_SQLSTATE
503,EXCHANGE
504,BETWEEN
```

9.2.4 将 MySQL 数据库中的表数据导入 HBase

1. 需要先创建 HBase 里面的表,再执行导入语句

```
hbase(main):001:0> create 'new_help_keyword','base_info'
0 row(s) in 3.6280 seconds
=> Hbase::Table - new_help_keyword
hbase(main):002:0>
```

2. 普通导入

```
[hadoop@hadoop3 ~]$ sqoop import \
--connect jdbc:mysql://hadoop1:3306/mysql \
--username root --password root \
--table help_keyword \
--hbase-table new_help_keyword \
--column-family person \
--hbase-row-key help_keyword_id
```

本 章 小 结

Apache Sqoop 被设计用于在一个 Hadoop 生态系统与 MySQL、Oracle、MsSQL、PostgreSQL 和 DB2 等关系数据库管理系统中的结构化数据存储之间传输数据。作为 Hadoop 生态系统不可或缺的一部分,Sqoop 启用了一个 MapReduce 作业(极其容错的分布式并行计算)来执行任务。Sqoop 的另一大优势是其传输大量结构化或半结构化数据的过程是完全自动化的。

参 考 文 献

[1] 徐郡明. Apache Kafka 源码剖析[M]. 北京:电子工业出版社,2017.
[2] 扎心了,老铁[EB/OL]. [2020-03-17]. https://www.cnblogs.com/qingyunzong/.
[3] ZooKeeper 教程[EB/OL]. [2020-04-12]. https://www.runoob.com/w3cnote/zookeeper-tutorial.html.